SKILL TRAINING
in MULTIMODAL
VIRTUAL
ENVIRONMENTS

Human Factors and Ergonomics

Series Editor

Gavriel Salvendy

Professor Emeritus
School of Industrial Engineering
Purdue University

Chair Professor & Head
Dept. of Industrial Engineering
Tsinghua Univ., P.R. China

PUBLISHED TITLES

Conceptual Foundations of Human Factors Measurement, *D. Meister*

Content Preparation Guidelines for the Web and Information Appliances:
Cross-Cultural Comparisons, *H. Liao, Y. Guo, A. Savoy, and G. Salvendy*

Cross-Cultural Design for IT Products and Services, *P. Rau, T. Plocher and Y. Choong*

Designing for Accessibility: A Business Guide to Countering Design Exclusion,
S. Keates

Handbook of Cognitive Task Design, *E. Hollnagel*

The Handbook of Data Mining, *N. Ye*

Handbook of Digital Human Modeling: Research for Applied Ergonomics
and Human Factors Engineering, *V. G. Duffy*

Handbook of Human Factors and Ergonomics in Health Care and Patient Safety,
Second Edition,
P. Carayon

Handbook of Human Factors in Web Design, Second Edition,
R. Proctor and K. Vu

Handbook of Occupational Safety and Health, *D. Koradecka*

Handbook of Standards and Guidelines in Ergonomics and Human Factors,
W. Karwowski

Handbook of Virtual Environments: Design, Implementation, and Applications,
K. Stanney

Handbook of Warnings, *M. Wogalter*

Human–Computer Interaction: Designing for Diverse Users and Domains,
A. Sears and J. A. Jacko

Human–Computer Interaction: Design Issues, Solutions, and Applications,
A. Sears and J. A. Jacko

Human–Computer Interaction: Development Process, *A. Sears and J. A. Jacko*

The Human–Computer Interaction Handbook: Fundamentals, Evolving
Technologies, and Emerging Applications, Second Edition,
A. Sears and J. A. Jacko

Human Factors in System Design, Development, and Testing,
D. Meister and T. Enderwick

Introduction to Human Factors and Ergonomics for Engineers, Second Edition,
M. R. Lehto

Macroergonomics: Theory, Methods and Applications, *H. Hendrick and B. Kleiner*

Practical Speech User Interface Design, *James R. Lewis*

Skill Training in Multimodal Virtual Environments, *M. Bergamsco,
B. Bardy, and D. Gopher*

PUBLISHED TITLES (CONTINUED)

Smart Clothing: Technology and Applications, *Gilsoo Cho*

Theories and Practice in Interaction Design, *S. Bagnara and G. Crampton-Smith*

The Universal Access Handbook, *C. Stephanidis*

Usability and Internationalization of Information Technology, *N. Aykin*

User Interfaces for All: Concepts, Methods, and Tools, *C. Stephanidis*

FORTHCOMING TITLES

Computer-Aided Anthropometry for Research and Design, *K. M. Robinette*

Foundations of Human–Computer and Human–Machine Systems, *G. Johannsen*

The Human–Computer Interaction Handbook: Fundamentals,
Evolving Technologies, and Emerging Applications, Third Edition, *J. A. Jacko*

Handbook of Virtual Environments: Design, Implementation, and Applications,
Second Edition,
K. S. Hale and K M. Stanney

The Science of Footwear, *R. S. Goonetilleke*

SKILL TRAINING *in* MULTIMODAL VIRTUAL ENVIRONMENTS

Edited by

MASSIMO BERGAMASCO
BENOÎT BARDY • DANIEL GOPHER

CRC Press
Taylor & Francis Group
Boca Raton London New York

CRC Press is an imprint of the
Taylor & Francis Group, an **informa** business

CRC Press
Taylor & Francis Group
6000 Broken Sound Parkway NW, Suite 300
Boca Raton, FL 33487-2742

First issued in paperback 2017

© 2013 by Taylor & Francis Group, LLC
CRC Press is an imprint of Taylor & Francis Group, an Informa business

No claim to original U.S. Government works

ISBN-13: 978-1-4398-7895-8 (hbk)
ISBN-13: 978-1-138-07548-1 (pbk)

Library of Congress Cataloging-in-Publication Data

Skill training in multimodal virtual environments / editors, Massimo Bergamasco,
    Benoit Bardy, Daniel Gopher.
        p. cm. -- (Human factors and ergonomics)
    Includes bibliographical references and index.
    ISBN 978-1-4398-7895-8 (hardback)
    1. Occupational training. 2. Virtual reality in education. I. Bergamasco, Massimo. II.
Bardy, Benoît G. III. Gopher, Daniel.

    HD5715.S595 2012
    658.3'124--dc23   2012021071

Visit the Taylor & Francis Web site at
http://www.taylorandfrancis.com

and the CRC Press Web site at
http://www.crcpress.com

Contents

SECTION I   Scientific Foundations

SECTION II   Engineering and Technology of Virtual Reality Training

SECTION III   Digital Representation and Modeling of Skill

SECTION IV   SKILLS Demonstration Platforms

SECTION V Propects of Multimodal Virtual Stimulators in the Training and Transfer of Skills

Preface

This book presents the scientific background, research outcomes, engineering developments, and evaluation studies conducted during the five years (2006 to 2011) of the project SKILLS (Multimodal Interfaces for Capturing and Transfer of Skills), funded by the European Commission under its Sixth Framework Programme for Research and Technological Development. The SKILLS project aimed at evaluating and exploiting robotics and virtual environments technologies for the training of specific skills, and it addressed skill acquisition according to a novel approach, setting aside the mainstream assumptions of common computer-assisted training simulators. The SKILLS approach was to generate new training scenarios in which the user can afford new experiences in the performance of the devised task.

The SKILLS Consortium includes universities and research centers with expertise in cognitive science, psychology, interaction design, virtual environments, augmented realities, and robotics, together with industries representative of the targeted application domains for novel multimodal training technologies.

Individual skills are difficult to describe. The modeling of a specific act still represents a challenge in cognitive science, and the same applies for the control of humanoid robots or for the replication of skilled behavior of avatars in the digital domain.

Research on training of human skills represents a state-of-the-art issue in the field of laparoscopic and computer-assisted surgery, where systems based on virtual environments and haptic technologies are receiving strong approval by the scientific community. Other examples of skill training are maintenance activities in the industrial field, where augmented reality technologies are currently used to assist the human operator in complex manipulative operations. The design of such training systems exploits virtual environment technologies to re-create the context of working and to realistically replicate the effects of specific actions of the operator. In contrast, current systems do not include capturing systems that describe the special skills of high-qualified persons while executing their challenging manipulations (e.g., to describe what makes a surgeon a good one). A true interpretation of the task and of the human action is still missing; therefore skill transfer is evaluated by external assessment. The SKILLS project has introduced a novel approach to skill capturing, transfer, and assessment based on enactive paradigms of interaction between the human operator and the interface system devoted to mimic task conditions, and it has extended the use of multimodal virtual environments for training to new fields, such as sport and entertainment.

The research in SKILLS addressed the design of new multimodal systems able to handle human skills. Fundamental aspects of skill analysis, including cognitive science and interaction design, were taken into consideration to identify guidelines for driving the analysis of the motion of skilled persons and to obtain a digital representation of human skill, as well as methodologies and techniques for capturing and rendering skills for training through digital technologies.

To be able to acquire, store, and transfer specific skills, such as those of a crafts-man, a surgeon, or an expert juggler into the digital domain can generate new ways of interacting with the computer and new ways of communicating knowledge through it.

The process of acquiring and storing skill allows for the generation of digital archives of performed acts which may be lost when the skilled person loses his or her capabilities due to illness or declining years.

In this context, specific (complex) tasks have been identified suitable to become the focus of training scenarios based on the application of robotics and virtual environments technologies. These tasks belong to different application domains—sports and entertainment, surgery and rehabilitation, industrial maintenance and assembly—and specific training scenarios and demonstrators (i.e., combinations of technological platforms and training protocols based on robotics and virtual environments) have been associated with domains such as rowing and juggling, maxillofacial surgery and upper limb rehabilitation, programming by demonstration in robotics, and maintenance in industry.

The concept of learning accelerators has been introduced to underline the importance of the process that customizes and exploits a subset of the multimodal data flows characterizing the interaction of the user with virtual environments to design a specific experience capable of improving and accelerating the acquisition of a specific skill.

This book illustrates the experience of the SKILLS project by describing the research scope and challenges and showing how they were tackled in SKILLS. Each chapter has been written by researchers who actively participated in the project and contributed to its most prominent results. The chapters are grouped into five sections. In Section I, the scientific foundations assumed in SKILLS for setting the scene of its research and defining the approach to the targeted challenges for multimodal training with Virtual Environments and augmented reality are introduced (Chapters 1–3). Section II consists of Chapters 4–8, which describe skill capturing and rendering as conducted in the project, and how haptic and auditory interfaces have been used for the purpose. Then, in Section III, Chapter 9 deals with the representation of human skill in the digital domain, and Chapter 10 discusses the management of data generated with training in Virtual Environments and for the evaluation of such training processes. In Section IV, the six case studies that exemplify the project approach to multimodal training in Virtual Environments through dedicated demonstration platforms are presented through Chapters 11–16. Finally, in Chapters 17 and 18, the prospects of multimodal virtual simulators in training and skill transfer are illustrated as they emerge from the SKILLS perspective, and in relation to the state of the art in this field.

Acknowledgments

This book is the outcome of the cooperative research experience carried out in the Integrated Project SKILLS 035005, supported by funding in the priority thematic area Information Society Technologies of the Sixth EU Framework Programme for Research and Technological Development (FP6).

The editors express their gratitude to the European Commission project officer for the SKILLS project, Philippe Gelin (DG/INFSO/E1 Language Technologies, Machine Translation) and the EC unit head for multimodal interfaces, Bernard Smith.

The editors also wish to thank the independent experts who assisted the commission in the periodic reviews of the project: Micah Murray (Université de Lausanne, Switzerland), Hong Tan (Purdue University, West Lafayette, Indiana), Roberta Klatzky (Carnegie Mellon University, Pittsburgh, Pennsylvania), Martin Buss (Technical University of Munich, Germany), Darwin Caldwell (Italian Institute of Technology, Genoa, Italy), Marc Ernst (Max Planck Institute for Biological Cybernetics, Tübingen), and Dongheui Lee (Technical University of Munich, Germany).

The research and technological development activities carried out in the framework of the SKILLS project result from the intensive collaborative effort of many people and their organizations:

Scuola Superiore Sant'Anna, Perceptual Robotics Laboratory (PERCRO), Pisa, Italy (www.percro.org)
Massimo Bergamasco (project coordinator)
Carlo Alberto Avizzano
Antonio Frisoli
Emanuele Ruffaldi
Franco Tecchia
Elisabetta Sani
Federico Vanni
Alessandro Filippeschi
Vittorio Lippi
Edoardo Sotgiu
Oscar Osvaldo Sandoval
Leonard Johard
Luigi Borelli
Sahar El Khoury
Caterina Procopio

Fraunhofer–Institut für Graphische Datenverarbeitung (IGD), Darmstadt, Germany (www.igd.fhg.de)
Uli Bockholt
Sabine Webel

Timo Engelke
Manuel Olbrich
Harald Wuest
Didier Stricker
Alain Pagani

Commissariat à l'énergie atomique et aux énergies alternatives (CEA), France
(www.cea.fr/)
Florian Gosselin
Sylvain Bouchigny
Christine Mégard
Claude Andriot
Fabien Ferlay
Moustapha Hafez
Florian Periquet
Florent Souvestre

Université Montpellier-1, Efficience et Deficience Motrices (UM-1), Montpellier,
France (www.edm.univ-montp1.fr/)
Benoît Bardy
Denis Mottet
Julien Lagarde
Sebastien Villard
Grégory Zelic
Manuel Varlet
Charles Hoffmann
Isabelle Laffont
Sofiane Ramdani
Didier Delignières

DLR, Institut für Robotik und Mechatronik, Wessling, Germany (www.dlr.de/rm/)
Carsten Preusche
Thomas Hulin
Katharina Hertkorn
Ulrich Hagn
Ulrich Seibold

Fundación Tecnalia Research & Innovation—Unidad de Siderurgia
(TECNALIA), Spain (www.tecnalia.com/)
Ana Rosa Carrillo
Teresa Gutierrez
Sara Casado

Centro de Estudios e Investigaciones Técnicas de Gipuzkoa, Departamento de Mecánica Aplicada (CEIT), San Sebastián, Spain (www.ceit.es/)
Emilio Sanchez
Jorge Rodriguez
Angel Suescun
Iker Aguinaga
Luis Unzueta
Jorge Juan Gil

Technion—Israel Institute of Technology (TECHNION), Haifa, Israel (www1.technion.ac.il/en)
Daniel Gopher
Eldad Yechiam
Maria Korman
Nirit Gavish
Vered Erev-Yehene
Guy Hochman
Arava Kallai
Stas Krupenia
Rotem Lammfromm
Dror Lev
Yfat Shorr
Ariel Telpaz
Yoni Tarlovsky
Geva Vashitz
Kinneret Weiss

University of Tampere, Department of Computer Sciences (UTA), Finland (www.cs.uta.fi/)
Grigori Evreinov
Roope Raisamo

Queen's University Belfast, School of Music and Sonic Arts (QUB), Northern Ireland, United Kingdom (www.sarc.qub.ac.uk/)
Sile O'Modhrain
Nick Gillian
Issartel Johann

Aalborg University, Acoustics Section (AAU), Aalborg, Denmark (www.es.aau.dk/sections/acoustics/)
Pablo Hoffmann
Dorte Hammershoi
Soren Krarup Olesen

Haption S.A. (HAPTION), France (www.haption.com/)
  Loic Tching
  Tony Chabiron
  Jerome Perret

KUKA Roboter GmbH (KUKA), Augsburg, Germany (www.kuka.com/)
  Uwe Zimmermann
  Volker Schmirgel
  Rainer Bischoff

OMG plc (OMG), Oxford, United Kingdom (www.omg3d.com)
  Paul Smyth
  Andrew Stoddart
  Cristian Canton
  Emli-Mari Nel
  Morne Pistorius

Sidel SpA (SIDEL), Parma, Italy (www.sidel.com/)
  Luigi Armani
  Matteo Peveri
  Marco Vescovi
  Francesca Pisu

A word of thanks to Federico Vanni and Elisabetta Sani for their help in editing the manuscript in its early versions.

Our gratitude goes also to the publisher, CRC Press, a Taylor & Francis company, and in particular to Cindy Renee Carelli, senior acquisitions editor, and Laurie Schlags, project coordinator, who took on this manuscript project on the publisher's side.

The Editors

Massimo Bergamasco is Full Professor of Theory of Machines and Mechanisms at Scuola Superiore Sant'Anna, Pisa, Italy, where he teaches Mechanics of Robots, Perception, and Virtual Environments. His research activity deals with the study and development of haptic interfaces for the control of the interaction between humans and Virtual Environments. His scientific activity includes more than 200 scientific papers published in journals and international conference proceedings. Bergamasco has been the coordinator of the ENACTIVE Network of Excellence (www.enactivenetwork.org) and the coordinator of the SKILLS Integrated Project (www.skills-ip.eu).

Benoît Bardy is Full Professor of Human Movement Science at Montpellier I University, Montpellier, France. His research is concerned with the dynamical approach to problems of coordination and control of action. Two major aims of his work are to investigate how the numerous muscles, joints, or segments that constitute the human body are coordinated so as to promote functional actions, and to identify the role of movement-based information in the control of action, with a specific interest in "real-world" activities. His current research focuses on coordination and multimodal control of human posture and orientation, egomotion, as well as reaching and grasping in real and virtual environments, using motion-capture, virtual reality, and robotic devices. Bardy is the author of more than 200 publications in peer-reviewed journals (e.g., *Journal of Experimental Psychology: Human Perception and Performance, Experimental Brain Research, Behavioral and Brain Sciences, Perception and Psychophysics*, etc.), books, and conference proceedings. He is currently involved in several international professional organizations, including the International Society for Ecological Psychology (board of directors), the French Society for Movement and Sport Sciences (past president), the American Academy of Kinesiology and Physical Education (2004 International Fellow), the European Network of Excellence Enactive Interfaces (sixth European Framework Programme for Research and Technological Development [FP], board of directors), and the European Integrated Project SKILLS (sixth FP, board of directors). He belongs to the Institut Universitaire de France and is the director of EuroMov, the new center for research, technological development, and innovation in movement sciences in France.

Daniel Gopher is Professor Emeritus of cognitive psychology and human factors engineering and Director of the Research Center for Work Safety and Human Engineering at the Technion—Israel Institute of Technology, Haifa, Israel. He is a Fellow of the Human Factors and Ergonomics Society and the International Ergonomics Association. His research is on topics of attention, mental workload, control processes, training of skills, and their application to aviation systems, medical environments, sports, and safety at work.

Contributors

Carlo Alberto Avizzano
Scuola Superiore Sant'Anna di Studi
    Universitari e di Perfezionamento
Perceptual Robotics Laboratory
    (PERCRO)
Pisa, Italy

Benoît G. Bardy
Movement to Health (M2H) Laboratory,
    EuroMov
Montpellier-1 University (UM-1)
Montpellier, France

Massimo Bergamasco
Scuola Superiore Sant'Anna di Studi
    Universitari e di Perfezionamento
Perceptual Robotics Laboratory
    (PERCRO)
Pisa, Italy

Uli Bockholt
Fraunhofer–Institut für Graphische
    Datenverarbeitung (IGD)
Darmstadt, Germany

Sylvain Bouchigny
Commissariat à l'énergie atomique et
    aux énergies alternatives (CEA)
Fontenay-aux-Roses, France

Cristian Canton-Ferrer
OMG plc (OMG)
Oxford, United Kingdom

Sara Casado
Departamento de Mecánica Aplicada
Centro de Estudios e Investigaciones
    Técnicas de Gipuzkoa (CEIT)
San Sebastián, Spain

Timo Engelke
Fraunhofer–Institut für Graphische
    Datenverarbeitung (IGD)
Darmstadt, Germany

Vered Erev
Technion—Israel Institute of
    Technology (TECHNION)
Haifa, Israel

Alessandro Filippeschi
Scuola Superiore Sant'Anna di Studi
    Universitari e di Perfezionamento
Perceptual Robotics Laboratory
    (PERCRO)
Pisa, Italy

Antonio Frisoli
Scuola Superiore Sant'Anna di Studi
    Universitari e di Perfezionamento
Perceptual Robotics Laboratory
    (PERCRO)
Pisa, Italy

Nirit Gavish
Technion—Israel Institute of
    Technology (TECHNION)
Haifa, Israel

Daniel Gopher
Technion—Israel Institute of
    Technology (TECHNION)
Haifa, Israel

Florian Gosselin
Commissariat à l'énergie atomique et
    aux énergies alternatives (CEA)
Fontenay-aux-Roses, France

Teresa Gutierrez
Fundación Tecnalia Research and
    Innovation
Unidad de Siderurgia (TECNALIA)
San Sebastián, Spain

Dorte Hammershøi
Acoustics Section (AAU)
Aalborg University
Aalborg, Denmark

Charles Hoffmann
Movement to Health (M2H), EuroMov
Montpellier-1 University (UM-1)
Montpellier, France

Pablo F. Hoffmann
Acoustics Section (AAU)
Aalborg University
Aalborg, Denmark

Thomas Hulin
DLR
Institut für Robotik und Mechatronik
Wessling, Germany

Maria Korman
Technion—Israel Institute of
    Technology (TECHNION)
Haifa, Israel

Isabelle Laffont
Département de Médecine Physique et
    de Réadaptation
CH&U Montpellier
Montpellier, France

Julien Lagarde
Movement to Health (M2H) Laboratory,
    EuroMov
Montpellier-1 University (UM-1)
Montpellier, France

Vittorio Lippi
Scuola Superiore Sant'Anna di Studi
    Universitari e di Perfezionamento
Perceptual Robotics Laboratory
    (PERCRO)
Pisa, Italy

Christine Mégard
Commissariat à l'énergie atomique et
    aux énergies alternatives (CEA)
Fontenay-aux-Roses, France

Denis Mottet
Movement to Health (M2H) Laboratory,
    EuroMov
Montpellier-1 University (UM-1)
Montpellier, France

Manuel Olbrich
Fraunhofer–Institut für Graphische
    Datenverarbeitung (IGD)
Darmstadt, Germany

Carsten Preusche
DLR
Institut für Robotik und Mechatronik
Wessling, Germany

Jorge Rodriguez
Departamento de Mecánica Aplicada
Centro de Estudios e Investigaciones
    Técnicas de Gipuzkoa (CEIT)
San Sebastián, Spain

Emanuele Ruffaldi
Scuola Superiore Sant'Anna di Studi
    Universitari e di Perfezionamento
Perceptual Robotics Laboratory
    (PERCRO)
Pisa, Italy

Emilio Sanchez
Departamento de Mecánica Aplicada
Centro de Estudios e Investigaciones
    Técnicas de Gipuzkoa (CEIT)
San Sebastián, Spain

Volker Schmirgel
KUKA Roboter GmbH (KUKA)
Augsburg, Germany

Paul Smyth
OMG plc (OMG)
Oxford, United Kingdom

Franco Tecchia
Scuola Superiore Sant'Anna di Studi
    Universitari e di Perfezionamento
Perceptual Robotics Laboratory
    (PERCRO)
Pisa, Italy

Ariel Telpaz
Technion—Israel Institute of
    Technology (TECHNION)
Haifa, Israel

Manuel Varlet
Movement to Health (M2H), EuroMov
Montpellier-1 University (UM-1)
Montpellier, France

Sabine Webel
Fraunhofer–Institut für Graphische
    Datenverarbeitung (IGD)
Darmstadt, Germany

Harald Wüst
Fraunhofer–Institut für Graphische
    Datenverarbeitung (IGD)
Darmstadt, Germany

Eldad Yechiam
Technion—Israel Institute of
    Technology (TECHNION)
Haifa, Israel

Gregory Zelic
Movement to Health (M2H), EuroMov
Montpellier-1 University (UM-1)
Montpellier, France

Uwe Zimmermann
KUKA Roboter GmbH (KUKA)
Augsburg, Germany

Section I

Scientific Foundations

1 Virtual Environments and Augmented Reality for Skill Training

Massimo Bergamasco

CONTENTS

INTRODUCTION

The project SKILLS deals with the use of Virtual Environments (VEs) and robotics for skill training. Skill acquisition and training is a complex process that involves motor, perceptual, and cognitive abilities as schematically depicted in Figure 1.1.

Through the use of VEs and robotics technologies, the aim of the SKILLS project is to allow the human operator to achieve better performance in the acquisition of the skill by focusing on a careful design of the interaction process between the human operator and the VE. As depicted in Figure 1.2, this is pursued by emphasizing proper optimization of the efferent-afferent flow of data in the interaction process.

The study of skill acquisition and transfer is becoming an important area of research not only in the framework of disciplines interested in the behavior of human beings, such as motor control, motor learning, ergonomics, and so forth, but also for disciplines such as robotics and VEs, in which the future trend of research is directed to the development of innovative ways for rendering robots and avatars to become more independent from the control of humans. It is plausible to think that in the coming decades, robots in the real environment and avatars in a parallel digital domain will interact with humans by directly cooperating in physical colocated operations.

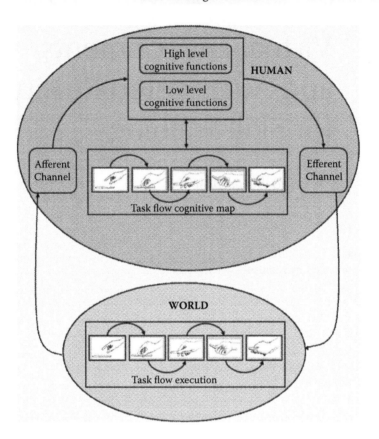

FIGURE 1.1 The cognitive-sensory-motor integration required by a skilled performance.

Our future artificial or synthetic partners will then have to possess a certain degree of skilled behavior.

Despite the long-term perspective of the vision described above, VE and robotics are exploited in the SKILLS project as new means to design specific training programs for specific tasks. The use of VE and robotics technologies in training represents a new paradigm in the framework of development of tools purposely designed to speed up the training process, with the intent to obtain better outcomes for the trainee.

It is difficult to find a suitable, coherent field of applications for VE, something for which VE can be the natural choice. Despite their inherent high technological threshold, there has not yet been a true effort in devising applications in which the true essence of VE can be fully recognized.

The conventional way of thinking about VE in terms of a single person interacting with a computer screen, or even inside a cave, by means of the tracked movements of his or her hands is still alive. In this paradigm, the goal is to try to translate in a VE setup what happens in a real environment and consequently design the interaction procedure and the behavior of virtual entities as a replica of something already seen in reality. Such an approach to the design of VE systems and applications is reductive because the use of the new media (the VE) is performed by exploiting

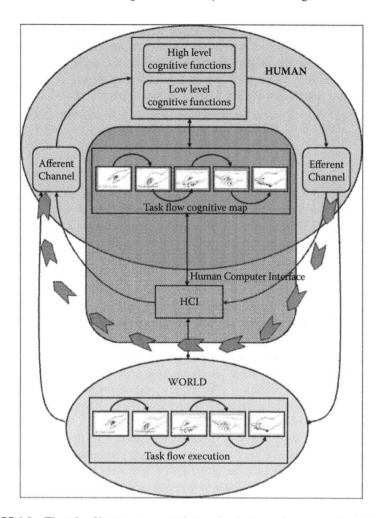

FIGURE 1.2 The role of human-computer interaction in the performance of a skill.

procedures (or paradigms) already existing in the old media, that is, interaction with the "real" environment.

The true essence of VE stands in different features that once properly considered represent the great potential of its technological advancement in innovative fields of application. What are these intrinsic features?

The development of VE systems and applications by having in mind the capability of replicating reality (i.e., a fine-grained engineering task aiming at copying the multi-sensory perception of virtual objects and interaction procedures and their effects on the human operator as in ordinary experience) is considered unlikely. Existing VE technologies for all the sensory modalities are at present extremely limited and unreliable for the generation of an adequate integrated experience of reality. The way in which humans perceive reality is already a simplification of reality by itself, so adding new sensory "noise" with the aim of providing alternative experiences is nonproductive.

Instead, the challenge of using VE for a specific functionality, such as skill acquisition, in a specific application domain (e.g., sports or surgery) must be considered according to an alternative methodology. The focus should not be that of replicating reality but generating experiences involving perceptual functions, cognitive functions, and motor functions that are adequate for the purpose.

According to this approach, the concept of "presence" in VE must be revised and truly exploited for the design of the appropriate technology.

PRESENCE IN VIRTUAL ENVIRONMENTS (VE)

Since the beginning of VE research in the late 1980s, the concept of presence has received strong attention. One of the first definitions from a pure engineering point of view was introduced by Sheridan (1992), who defined three main determinants for the sense of presence for teleoperators and VE systems. These are:

1. The amount of sensory information that the system is able to convey to the human operator. The greater the amount, the greater is the sense of presence in the artificial environment (or real environment in the case of a teleoperator). For example, for the visual sensory pathway implemented with a video screen presenting an image acquired by a black-and-white (b/w) camera in the remote environment, the generated sense of presence results in a reduced sense of "being there" with respect to the case in which the camera would have color capabilities. The same applies for stereo audio information with respect to binaural audio or for on/off haptic effectors with respect to fine-grained texture reproduction of haptic contact.
2. The capability of being able to move the sensors in the environment. When the user is able to visually explore the environment by exploiting the movement of the eyes, certainly his or her perception of the environment is augmented with respect to the case in which a static image would be exploited as taken from the same point of view and with the same perspective. Binaural audio intrinsically exploits the mobility of the head to acoustically acquire spatial geometry of the environment. The active movements of the fingertips allow the haptic system to implement exploratory procedures and acquire the texture of the contacted surfaces.
3. Control of the environment. This determinant refers to the ability to grasp objects and modify their relative position and orientation in space. The effect of being able to interact with the environment is extremely important to elicit a sense of presence: constraints or the inability to reach and grasp objects generate a sense of inadequacy in the framework of a specific task.

Sheridan suggested that when the three determinants can be ideally brought to their maximum values, a true sense of presence can be achieved by assuming that such a value can correspond to the "ideal" condition of presence as perceived in ordinary reality. However, present VE technologies do not allow such an ideal value to be reached in the common interaction procedures with VE, the main motivation being the high technological threshold of the different available technologies, in particular those related to the third determinants, such as haptics or robotics systems.

A more recent and deeper analysis on the sense of presence in VE has been conducted by Slater (2009) by taking inspiration from the theory of sensory-motor contingencies by O'Regan and Noë (2001a, 2001b). Slater introduced two orthogonal components that allow participants to generate a realistic response inside immersive VE.

The first component is called *place illusion* (PI) or the quale of having the sensation of being in a real place (Slater, 2009). The second, called *plausibility illusion* (Psi) refers to the illusion that the scenario being depicted in VE is actually occurring (Slater, 2009).

Sensorimotor contingencies (SCs) can be considered as the relations between the actions performed by the user and the associated variations in sensory inputs (i.e., actions we carry out in order to perceive). A VE system is able to support specific SCs that contribute to the perception of the environment.

A VE system that implements SCs similar to those utilized in a real environment can generate to the user the illusion of being located in the VE or, as described in several examples of the scientific literature, the "sense of being there" (Held and Durlach, 1992).

The term PI is related to such a sense of being there. On the other hand, Psi relates to "what is perceived, or better, the illusion that what is apparently happening is really happening (even though you know for sure that it is not)" (Slater, 2009).

EXPERIENCES IN VE

The true motivation for dealing with SCs is being able to understand the meaning of the "experience" generated in a VE while performing a task or carrying out a training procedure. As described by O'Regan and Noë, "Experience is not generated by brain processes themselves, but rather is constituted by the way these brain processes enable a particular form of 'give-and-take' between the perceiver and the environment" (O'Regan and Noë, 2001a). Experience in VE can then be defined as a temporally extended pattern of integrated activity, a sort of give-and-take between the user and the VE. Consequently, in order to achieve a significant experience of effective training in VE, it will be of paramount importance to focus on the essential aspects of the interaction (give-and-take or perception-action loops) paradigms between the human operator and the VE.

Our experience of the activities we carry out while doing or learning even a simple task originates from a complex integration of sensory, motor, and cognitive functions. The perceptual experience (i.e., the perceptual world in which we are working) is created by our senses and central nervous system (CNS) and can be considered as a functional representation of the physical world. The world of which we are perceptually aware is called the *phenomenal world*, which is distinct from what we call the *physical world*. Such a distinction is supported by the scientific literature on cognitive sciences (Loomis, 1992) and rehabilitation (Johnstone and Stonnington, 2009).

When interaction with VE occurs, the human operator should be put in the condition of being able to identify the contents of his or her experience as belonging to the external virtual world to which he or she is exposed. In other words, the phenomenal world in the two conditions of "real" and virtual experiences should be comparable. How can VE technologies, in terms of capturing, modeling, and rendering components, provide such a condition?

Instead of pushing the limits of technology in order to achieve realism of perceptual representation, our first assumption in SKILLS is focused on designing VE systems able to provide to the user a strong coherence between efferent data (commands to the muscles that are the effectors of the intended action) and afferent data (multimodal sensory feedback provided to the user by the VE interface system) (Loomis, 1992).

The coherent relationship between efferent and afferent data in a VE system as a determinant for achievement of a particular experience has been pointed out also by several authors involved in the study of perception (Epstein et al., 1986; Loomis, 1992; White, 1970).

When designing a specific VE interface system for a particular task, two aspects are essential in order to achieve coherency between efferent and afferent channels of the multimodal interface:

1. Transparency (i.e., the condition for the human operator to perform the task in VE without perceiving the effect of presence of the mediating system generating the appropriate stimuli). As Polanyi (1970) pointed out, the condition of transparency can be effective when "subsidiary awareness" of the mediating chain becomes almost null and, in this condition, the "focal awareness" by the user is maximized to the learning process of a skill acquisition (Loomis, 1992).
2. Fidelity (i.e., the ability of the VE interface system to generate appropriate stimuli without being affected by disturbing effects during time). The replication of specific interaction experiences should be guaranteed within a specific tolerance range during the execution of multiple training procedures.

The general design of a VE system for training purposes ought to start from the analysis of the perceptual experience to be generated and then proceed to the level of detail proper to the specific target task or application.

PERCEPTION IN VE

The involvement of a human operator in the execution of a particular task and his or her capability to learn the specific components of the skill and to maintain control of the task despite external perturbations belong to a perceptual "flow" in which all components of a multimodal VE system contribute according to a bilateral exchange of information.

If we analyze perceptual processes for each sensory modality, the integration of all stimuli into a coherent "flow," in which attention and memory functions also contribute, appears as a dynamic property of a system for which complexity is extremely high.

In fact, by considering only the main three sensory modalities, usually integrated within an immersive VE (i.e., vision, audition, and haptics), the amount of data and the functionalities to which they contribute are astonishing.

VISION

Visual perception covers a large number of perceptual primitives. Cortical visual pathways can be subdivided into two putative "where" (dorsal) and "what" (ventral) streams both generating from the early visual areas. The "what" stream is deputed to the analysis of form and color, while the "where" stream is related to the analysis of motion and spatial relations.

From spatial vision of the early level, other primitives deal with perception of colors, perception of space, and perception of motion. Middle vision brings the perceptual functionalities to the perception and recognition of objects, to attention and perception and understanding of scenes.

Head-mounted displays allow complete visual involvement of the user in the represented scenario, but the visual field represented is still partial with respect to that of the user. Consequently, the VE systems in SKILLS have exploited different solutions of stereo visualization systems composed of screens usually arranged for obtaining adequate co-location conditions with haptic feedback systems.

HEARING

The effective value of this modality is usually underestimated in current VE systems. However, hearing in the environment, in particular through binaural systems, is extremely important for the localization of sound in three-dimensional (3D) spaces, for the auditory analysis of scenes and also for music and speech perception. In SKILLS, the effect of adding sound feedback in the framework of specific operations in VE, such as in rowing or in surgery training systems, has proved to be valid to generate added cues to the human operator during the execution of specific tasks.

HAPTICS

Haptics is a perceptual system that exploits tactile and kinesthetic stimuli to acquire information from contact conditions of the user with the external environment. Kinesthesis together with the sense of balance detected by the vestibular system contribute to the proprioception perceptual system, which in turn, together with the sense of touch, generate somatosensation.

All the above components contribute to the sense of presence in VE because they allow for implementation of the determinant of the control of the environment. Without having the possibility to touch, explore, or grasp virtual objects, the component of the control of the environment could not be effective. Then, at present, haptics and, in general, somatosensation, are considered essential for the control of a true interaction with VE.

As visual perception, haptic perception can be referred to two primary functionalities: the "what" stream which is involved in perceiving objects and their properties, and the "where" stream deputed to the localization of objects with respect to a user's frame of reference. Haptic perception is intrinsically based on the perception-action loop, because it necessarily involves voluntary action to extract information from the external environment. Lederman and Klatzky (1987) have pointed out the correlation between the performed pattern of hand movements and the information the user intends to obtain from the environment. Exploratory procedures are executed by the hand in order to acquire information on texture, shape, hardness, temperature, or pain conditions from the explored objects. The same considerations apply in VE, where haptic interfaces represent the component of a VE system for allowing the user to perform exploratory or manipulative operations in VE. In SKILLS haptic interfaces represent an interaction modality largely utilized in almost all of the training scenarios.

CONSTRAINTS AND ACCELERATORS

Inside an operative scenario the human operator has to control specific actions while interacting with the external environment or other subjects; moreover, the task possesses specific procedures to be respected. A classic example is the operating room where the surgeon has to follow a specific operating procedure and the task presents critical conditions to be tackled and solved quickly.

The coordination and control of a task by the human operator are then subjected to different kinds of constraints:

1. Biomechanical and practical level of performance of the human operator. Biomechanical constraints refer to the reachable and dexterous workspace of the human limbs, and consequently of the possible reachable configurations that hands and arms can achieve when involved in a specific task. The level of performance in terms of force, fatigue, attentional effort, and so forth, can vary largely in different conditions.
2. Environmental constraints are those imposed by the conditions of the space in which the human operator performs the task. They include not only the geometry of the physical space but sometimes also sensory stimuli acting as disturbances during the execution of the task.
3. Task constraints are related to the execution of the task and include all those factors that drive its protocol or procedure, or the specific tool the user is obliged to use.

The design of training procedures in VE must consequently take into account the capability of generating or deleting specific virtual constraints during the execution of the task.

During the execution of a task, the control of the operation by the user in the framework of a specific environment refers to a perception-action flow that can be affected by changes in environmental or other types of constraints. The ability of the user is then that of being able to manage such variations and newly setting up perceptual-action procedures until the completion of the task. Perceptual feedback streams are intrinsically provided by multimodal sensory stimuli. However, improvements in execution can be obtained by generating appropriate feedback stimuli to the user in order to provide him or her with an indication of the level of performance achieved so far (knowledge of results and knowledge of performance).

In SKILLS, VE systems have also been exploited to generate the accelerators, which can be considered as selected perception-action loops for improving specific user's performances. The highly complex interaction loop during the execution of a task can be reduced by introducing a sort of specific, narrow band perceptual-action flow field, which can be tuned into the user's ability, and which focuses or directs the training procedures along a specified direction.

Accelerators have been introduced in SKILLS for specific components (either sensorimotor or cognitive) of the skill to be trained.

DIGITAL REPRESENTATION OF SKILL

The introduction of accelerators in the framework of the training process in VE brings the operative conditions to achieve a narrowed, although purposely selected, dimension of the task description which is in fact reduced with respect to the real conditions where the task will be performed. The introduction of accelerators can be compared to a metaphorical description of the task.

Moreover, as it happens for the large number of present VE systems, the behavioral component of the VE, which is devoted to model, generate, and control the behavior of the virtual entities represented in the VE and also the interaction with the human operator, is completely unaware of how the task has been defined or conducted. Usually the training protocol is defined externally (by an external entity, such as a coach or a trainer) and in this condition, the VE system only acquires data from the different interfaces (afferent channel to the human operator) and provides commands to the interface actuators (efferent channel from the human operator). In this typical way of operating, the VE behavioral component is limited to acquiring data and consequently alters the context of interaction between the VE system and the human operator.

However, although the introduction of constraints and accelerators can be considered an innovative component in the framework of present VE systems, in SKILLS a further effort has been made in order to provide in the VE system a certain level of awareness of the task and how it is performed by the human operator.

To this purpose a theoretical framework, called *digital representation of skill*, has been studied and developed: the digital representation allows us to intrinsically understand and model the relationship among the different dynamical constraints generated in the perception-action flow during the execution of the training procedure and consequently drive the VE system to adjust the operative conditions in order to achieve better training performance.

By means of digital representation of skill, the VE systems developed in SKILLS and utilized in the different application scenarios have achieved the following functionalities:

1. Awareness in the benchmarking of the task, which makes the training system able to answer the question "how well has the task been performed?" At present the measure of performance is determined by external measures (e.g., by evaluating the execution time with a chronograph). Having the digital representation implemented, the VE-based training system is not only able to provide an indication of how the human operator is performing but also (a) to establish statistical evaluation of the accuracy by which the task is performed (i.e., to define the probabilities of error during the execution) and (b) the eventual differences of the actual and planned constraints.
2. Generation of autonomous characters (i.e., either robots or avatars) that (a) are able to learn from the demonstration of the task performed by a human operator and (b) can be respondent to variations to the environmental constraints.
3. Tailoring the training procedure for a specific human subject. The digital representation can provide both intrinsic and extrinsic specific feedback stimuli to the human operator according to his or her physiological constraints and performances and level of practice.

CONCLUSIONS

An overview of the rationale on which the SKILLS project has been conceived and carried out has been presented. As depicted in Figure 1.3, conventional training is carried out according to a serial methodology including the definition of training protocols, followed by practice and assessment of performance. The innovative aspects of SKILLS VE technologies and systems applied to training refer to the introduction of the accelerator components applied to specific subskills (i.e., components of the skill belonging to the task to be trained), as depicted in Figure 1.4. A

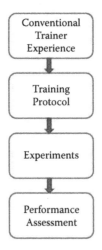

FIGURE 1.3   The serial methodology of conventional training.

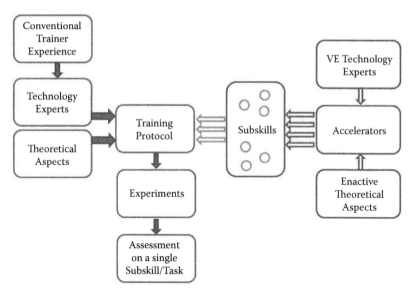

FIGURE 1.4   Innovative SKILLS training virtual environment systems based on accelerators.

further innovation is brought by the digital representation of skill which can be used in different application domains for interpreting expert behaviors and modeling the relevant components of the underlying skills.

REFERENCES

Epstein, W., Hughes, B. Schneider, S., and Bach-y-Rita, P. 1986. Is there anything out there? A study of distal attribution in response to vibrotactile stimulation. *Perception*, 15(3): 275–284. London, UK: Pion.

Held, R.M., and Durlach, N.I. 1992. Telepresence. *Presence: Teleoperators and Virtual Environments*, 1: 109–112.

Johnstone, B., and Stonnington, H.H. (eds.). 2009. *Rehabilitation of Neuropsychological Disorders: A Practical Guide for Rehabilitation Professionals*. London: Psychology Press.

Lederman, S., and Klatzky, R. 1987. Hand movements: A window into haptic object recognition. *Cognitive Psychology*, 19: 342–368.

Loomis, J. 1992. Distal attribution and presence. *Presence: Teleoperators and Virtual Environments*, 1: 113–119.

O'Regan, J., and Noë, A. 2001a. A sensorimotor account of vision and visual consciousness. *Behavioral and Brain Sciences*, 24: 939–972.

O'Regan, J., and Noë, A. 2001b. What it is like to see: A sensorimotor theory of perceptual experience. *Synthese*, 129: 79–103.

Polanyi, M. 1970. What is a painting? *British Journal of Aesthetics*, 10: 225–236.

Sheridan, T. 1992. Musings on telepresence and virtual presence. *Presence: Teleoperators and Virtual Environments*, 1: 120–126.

Slater, M. 2009. Place illusion and plausibility can lead to realistic behaviour in immersive virtual environments. *Philosophical Transactions of the Royal Society B: Biological Sciences*, 364(1535): 3549–3557.

White, B. 1970. Perceptual findings with the vision-substitution system. *IEEE Transactions on Man-Machine Systems*, 11: 54–59.

2 Development of Training Platforms in Multimodal Virtual Reality Environments

Daniel Gopher

CONTENTS

INTRODUCTION

Multimodal, immersive, virtual reality (VR) techniques open new perspectives for perceptual-motor skill trainers. They also introduce new risks and dangers. This chapter introduces and discusses the building blocks, benefits, and pitfalls of developing multimodal VR training environments for perceptual motor skills.

Training simulators for complex tasks are being used in increased frequency since the end of World War II, which also marks the beginning of the technological age revolution. With the growing complexity of systems and their operational environments, the required duration of training, and the increased costs of errors, on-the-job practice became difficult or impossible, and alternative training and simulation environments have been developed to enable skill acquisition and learning. Flying, driving, space missions, surgery, power plant, and process control jobs are salient examples for tasks, skills, and competencies for which training simulators have been developed. Early simulators were physical and mechanical analogues of their represented systems. With the advance of computer technology, simulators have become more and more hybrid. System dynamics, visual field of view, and audition have been increasingly driven and generated by computers. Contemporary developments in sensors and display capabilities and the exponential increase in computation speed and storage capacity led the way to the development of multimodal virtual environments. In these environments the operator is immersed, experiences multimodal sensations, and interacts with virtual objects including other humans (Riva, 2006). Visual and auditory have been in the study and design of simulators from their inception. A new and important addition is the inclusion of haptics: the ability to feel and exercise force, touch, texture, and kinematics. Haptic technology is developing rapidly, and haptic interfaces are now being incorporated in many virtual worlds. It is hence a quite conservative expectation that the multimodal virtual reality platforms will dominate the next generation of training simulators. From the vantage points of training and motor and cognitive sciences, this development carries with it some exciting prospects and serious challenges. These are the focal topics of this chapter. Their study has been the prime objective of the research, engineering, and modeling work that has been conducted in the SKILLS project and described in this book.

At the outset of this chapter it is important to emphasize that the topics of training and transfer of skills have not been the focus of interest and effort in contemporary research and development of virtual environments. When studying interfaces, human behavior, task performance, and interaction in virtual environments, the focus has been on the fundamentals of developing "virtualization" and the compelling concepts of "presence" and the "immersion" experience of the involved human performers. Sanchez-Vives and Slater proposed the following definitions for the three concepts: *Virtualization*: "the process by which a human viewer interprets a patterned sensory impression to be an extended object in an environment other than that in which it physically exists" (2005, p. 332). The interest is hence in the properties and dimension that should be captured and represented in order to create virtual objects. *Presence*: "The common view is that presence is the sense of being in a VE rather than the place in which the participant's body is actually located." A fundamentally different view is that presence is "tantamount to successfully supported action in the environment." The argument is that reality is formed through actions rather than through mental filters, and that "the reality of experience is defined relative to functionality, rather than to appearances" (2005, p. 333). *Immersion*: "A person is immersed in an environment that is realized through computer-controlled display systems, and might be able to effect changes in that environment" (2005, p, 332).

One aspect in the evaluation of virtual environments is subjective feeling of "presence" and "immersion," which is measured by a variety of rating scales. A second approach to evaluation is by the use of behavioral and performance measures, taken during relatively short duration interactions with the VR environment. The main interest is the level and correspondence between behavior in the real and virtual environments. Sanchez-Vives and Slater (2005) further argue that there is a need to examine presence and immersions by employing neurophysiological and imaging measure and comparing them to brain activations of the corresponding real-life task.

When evaluating the value and possible contribution of VR technologies to the training of skills, presence and immersion are related but are not the prime focus. The three key constructs for evaluating the value of a training system are *relevance, facilitation,* and *transferability. Relevance* refers to system ability to provide the relevant experience for the development of competency and skill on a task. *Facilitation* refers to the inclusion of facilitation and guidance to assist and accelerate the acquisition of the designated skill. *Transferability* emphasizes the transfer of training and skill levels from VR training to task performance in the real world. These three constructs, their requirements, challenges, and the crucial criteria for the evaluation of training systems are very different and distinguished from those of studying presence and testing immersion. High levels of presence and immersion may be important motivators and acquisition augmenters, if there is a good match (fidelity) between the VR and real-life task experience. However, in case of a mismatch or diversion, they may be harmful. High subjective sense of presence formed in spite of experiential diversions from reality may lead to illusionary conjunctions and to reduced or even negative transfer. In contrast, relevant training experience may enhance skills and improve transfer even when presence and immersions are low or absent (think for example of finger dexterity exercises in piano or aerobic workouts for basketball players). When developing training simulators, it is sometimes necessary to create deliberate diversions and reduce physical fidelity to avoid the emergence of involuntary and unconscious illusionary conjunctions and underline the important principles to be extracted and acquired. The study of the value and prospects of VR technologies for skills training should therefore establish and follow its own track, and focus on topics directly related to the acquisition and transfer of skills.

SIGNIFICANCE OF MULTIMODAL EXPERIENCE

For humans and most organisms, potential sensory stimulation is characterized by simultaneous changes in multiple forms of ambient energy (Stein & Meredith, 1993; Stoffregen & Bardy, 2001). Ernst and Bülthoff (2004) suggested that no single sensory signal can provide reliable information about the three-dimensional structure of the environment in all circumstances, a suggestion that has received psychophysical validations in the context of distance perception (Mantel et al., 2005). Synergy (the merging information from multiple aspects of the same task), redundancy, and increased bandwidth of information transfer are proposed benefits of multimodal presentation (Sarter, 2006). If information is presented in redundant multiple modalities, input in various concurrent forms and different aspects of the same process is presented. When the varieties of multimodal cues are registered by the sense organs, the information must be integrated by neural structures to form a coherent and holistic representation of the object. It appears that our perceptual-motor abilities follow from cross-modal processing of our surroundings, and that a coherent and unified percept can emerge from cross-modal binding of different sensory cues with varying amounts of reliability (Lalanne & Lorenceau, 2004). For training purposes it is important to note that presenting information in one sensory modality (such as sound) can immediately draw a person's attention to other sensory stimuli presented in the same location (such as vision and tactile sensations) (Spence et al., 2004).

Comparative analysis of the interaction patterns involved in redundant and non-redundant cue processing provides evidence for the robustness of the principle of cross-modal neural synergy that applies regardless of the stimulus content. In addition, a comparative analysis provides evidence for the high flexibility of the neural networks of integration that are sensitive both to the nature of the perceptual task and to the sensory skill of the individual in that particular task (Fort et al., 2002). Interestingly, perceptual phenomena associated with single sensory channel processes (such as change blindness and change deafness) can also occur when multiple modalities are presented (Auvray et al., 2008). Several studies have shown that using a combination of two modalities—visual, auditory, or haptic—increases performance compared to using each of the modalities separately (Doyle & Snowden, 2001, Murray et al., 2005). A trimodal combination of auditory, visual, and haptic stimuli was detected faster than the shortest of the bimodal combination (Diederich & Colonius, 2004).

Nevertheless, it is important to note that multimodal experience is not equivalent to the sum of modalities' experience. Vision is often considered the prime and preferred sensory modality for humans, and as such it can dominate the experience. Burr and Alais (2006) suggested that visual information does not simply dominate over auditory information, but how multimodal information is combined depends on the reliability of the stimulus inputs, and the more reliable input often dominates. Bresciani, Dammeier, and Ernst (2008) suggested that vision alone plays a minor role in feeling contact with objects, at least when touch and sound are available, and that audition could be used to enhance the feeling of contact if it is appropriately coupled with touch.

TRAINING IN MULTIMODAL ENVIRONMENTS

Multimodal displays support flexible efficient communication, are easy to learn, can be used in challenging situations, and people enjoy using them (Oviatt, 2002). For example, Tindall-Ford, Chandler, and Sweller (1997) found that participants who studied materials incorporating audio text and visual diagrams or tables performed better than those who studied in a conventional, visual-only format. As per the previous discussion on multimodal integration, it appears that multimodality can enhance skill acquisition. Nevertheless, using multimodality displays for training is not straightforward. For example, the dominance of visual feedback in simple motor tasks, discussed above, was reflected in training studies. Training under conditions of combined visual and proprioceptive knowledge of results reduced the efficiency of proprioceptive feedback in a later test during which proprioception was the only available feedback source. This finding implies that non-visual knowledge of results is suppressed or not employed when visual knowledge of results is available, with participants learning to rely solely on visual knowledge of results (e.g., Adams et al., 1977; Yechiam & Gopher, 2003). Virtual spatial sounds can also be used in multimodal contexts to supplement visual information in searching tasks (Bolia et al., 1999).

Advantages of Training in Simulated and Virtual Environments

Simulated and virtual environments are being increasingly used for teaching and training in a range of domains including surgery (Howell et al., 2008), aviation (Blake, 1996), anesthesia (Gaba, 2005), rehabilitation (Holden, 2005), and driving (Godley et al., 2002). These simulated and virtual worlds can be used for acquiring new skills or improving existing ones (Weller, 2004). The usefulness of simulators for training was demonstrated in several studies. Chopra et al. (1994) demonstrated that training on anesthesia simulators improved the acquisition and retention of knowledge as compared to receiving the same material via lectures. A review by Holden (2005) on the current methods of VR applications for physical rehabilitation identified that in most cases movements learned in a virtual environment transfer to equivalent real-world tasks and in some cases to other untrained tasks. Together, the research reviewed above demonstrates the value of and need for successful and efficient training simulators. Virtual reality thus appears a worthwhile tool for teaching and learning.

Multimodal interfaces succeed in creating a stronger sense of presence by better mimicking reality (Romano & Brna, 2001). The sensorial richness of multimodal environments translates into a more complete and coherent experience of the virtual world, and therefore the sense of being present in the virtual realm is stronger (Witmer & Singer, 1998). The experience of being present is especially strong if the virtual world includes haptic (tactile and kinesthetic) sensations (Reiner, 2004). Additionally, navigation in VEs represents a type of spatial task in which performance can be enhanced by employing spatial sound (Gonot et al., 2006). One reason for the increase in subjective fidelity when being in a multimodal virtual environment originates from the congruency between perceptual modalities (i.e., a certain pattern of stimulation within the global array, which resembles that occurring in the real situation) (e.g., Stoffregen & Bardy, 2001). However, one has to be careful with the concepts of presence and fidelity when interacting with a multimodal interface. Metrics for fidelity fall into two classes: fidelity of subjective experience (presence, or experiential fidelity, or subjective fidelity) or fidelity of performance (performance fidelity, or action fidelity, or functional fidelity) (Stoffregen et al., 2003). While subjective fidelity is usually operationalized in terms of conscious reports such as questionnaires or numerical intensity ratings, action fidelity is defined in terms of relations between performance in the simulator and performance in the simulated system. Action fidelity exists when performance in the simulator transfers to behavior in the simulated system. Metrics of action fidelity are more useful as constraints on the design and evaluation of training and skill acquisition than metrics of subjective fidelity.

Risks of Using Multimodal Virtual Reality and Augmented Reality Feedback in Training

If fidelity cannot be preserved or is hard to achieve, it is much better to avoid the use of the VR instantiation, or alternatively, particular care must be taken to develop a training program that identifies the VR and real-world mismatch to the trainee and provides compensatory training mechanisms. For example, the standard view of the U.S. Air Force is that platform motion is not recommended in flight simulators for centerline thrust aircraft (Cardullo, 1991), because for military planes performing large and fast maneuvers, the motion errors in the moving base simulators and those of actual flight are considerable and result in negative transfer because of illusionary conjunctions; therefore motion platforms are rarely used (Kaiser & Schroeder, 2003). Under these conditions, fixed-base simulators are preferred, because the distinction between the simulator and the world is clear and easy to observe. Additionally, the artificially constrained field-of-regard (the entire visual field available for sampling) found in most high-fidelity flight simulators can cause pilots to adopt novel visual strategies particular to the simulator (Kaiser & Schroeder, 2003). The novel viewing strategies are unlikely to transfer to the real world; therefore, real-world performance will not match VR performance.

Impoverished environments can be created to emphasize certain components of the virtual world, intended to overcome the dominance of one modality over another, and to help the trainee to develop certain sensitivities, capabilities, and modes of behavior that are otherwise suppressed in real-life operation conditions. For example, when using the standard computer keyboard, natural dominance of vision causes trainees to intuitively adopt a visually guided typing strategy. Acquisition of touch-typing skills, based on proprioceptive information feedback from the hand and fingers, does not develop without deliberate, long, and tedious training (Wichter et al., 1997). A secondary task paradigm, in which the visually guided typing strategy was made less attractive, led to a faster acquisition of touch-typing skills and higher performance levels both at the end of training and in the retention tests (Yechiam & Gopher, 2003).

An important advantage of the VR technology is the ease of providing augmented sensory feedback, visual guidance, auditory directors, or augmented haptic cues. However, the greatest potential danger of VR and augmented reality (AR) applications is that performers become increasingly dependent on features of the VR devices, which may inhibit the ability to perform the task in the AR-feature's absence or in the case of technology failure. For example, during laparoscopic surgeries, surgeons may be required to discontinue the procedure and move to traditional, manual surgery. Retention of manual surgery capabilities and the switch between procedures are not trivial. To overcome skill lapses associated with the potential switching from laparoscopic to manual surgery, training programs should include as few as possible AR features that would not be present in the actual operational environment, or alternatively include training phases that reduce the dependence of these features. An alternative example comes from aviation when multiple VR systems are used. Pilots are trained in simulators and on actual missions to land their airplanes using direct vision, without instruments or an instrument landing system. Developing dependence, or at least reliance, on VR features that do not exist in the real environment or are very different from their real-world counterparts (e.g., airplane motion cues) can result in negative transfer to the real world.

DEVELOPING TRAINING IN MULTIMODAL ENVIRONMENT

Training platforms are engineering systems developed to enable the acquisition of targeted skills and direct their development under guided instruction and training. Skill is defined as a well-organized knowledge base in long-term memory, developed with experience and training for the performance of a targeted task (e.g., Anderson, 1981, 1982; Chase & Ericsson, 1981; Meyer & Kieras, 1997). The nature, form, and format of the stored and memorized information may vary widely. It may be semantic, motor, modality specific, or abstract. It may be episodic declarative or procedural. In all cases its main purpose and value is that experience and knowledge of the past can help to cope with future and new demands. This is the most general and inclusive meaning of the term *skill*. Skill is hence best tested by its retention and transfer to more efficient coping with recurring or new events. Task performance levels alone may not be a sufficient indication for learning, as they may be the product and representation of imitating, copying, or following instructions. Learning requires active interaction and control and is maximized when it results from intentional efforts (Gopher, 2006, 2007; Schmidt & Bjork, 1992).

The five main building blocks of a training platform and a training program are as follows:

1. *Task specification*: A clear specification of the task to be learned, the skills to be acquired, the objectives of training, and the designated criteria of graduation. When developing a training platform for juggling, rowing, maxillofacial surgery, or upper limb rehabilitation, what components, compatibilities, and knowhow should be acquired? What objectives have to be achieved?
2. *Training program*: Design of task scenarios, task versions, and difficulty manipulation that best represent the skill components, training objectives, typical encounters, and key requirements of the task. A richer and diverse training environment affords the development of a more flexible and higher-level competence (Bjork and Schmidt, 1992; Gopher, 2007).
3. *Performance measures*: Identification of key response and performance measures as well as progress criteria to evaluate trainee progress on relevant aspects of task performance and enhanced competence.
4. *Feedback and knowledge of results*: Definition of desired feedback indices and knowledge of results information to be given to trainees, as well as their frequency and mode of presentation.
5. *Transfer of training*: When training is conducted as a preparatory stage, in a separate environment, or in a simulator, the fifth important consideration is the transfer of training. The evaluation and testing of the relevance of the training experience in the learning environment, by examining the level of transfer from the training platform to actual task performance.

Each of these five components has to be dealt with and presents major scientific and engineering challenges in the development of any training environment. More specifically, within the SKILLS project we attempted to develop exemplary platforms in the three fields of sport and entertainment (rowing, juggling), medical and rehabilitation systems (maxillofacial surgery, upper limb rehabilitation), and industrial work (industrial maintenance and assembly, programming by demonstration). A detailed discussion of the scientific and engineering efforts associated with the above five topics in each of the developed platforms is beyond the scope of this introduction chapter. It is described in Chapters 11 through 16, each of which focuses on the work conducted with one training platform. This introductory chapter highlights general considerations that were common to all developed platforms.

TASK ANALYSIS

When developing a training platform for any task, the most important primary topic and challenge is the exact specification of the task to be trained, the objectives to be achieved, and the competencies to be focused upon. This work draws upon the ensemble of task analysis methodologies that have been developed in human factors engineering (Salvendy, 1997: Shneiderman, 1998). Task analysis is a structured process for analyzing the ways that people interact with systems, their actions and interactions with the system, the task demands, and the operator's requirements of the system. The main goal of the task analysis is to specify the requirements for user-centered engineering design and human system interfaces development of training protocols. Task analysis is used to identify and define all the operations needed in order to achieve each goal and target of the task, and perform all subtasks as well as the basic relationships between the tasks and subtasks. After performing the analysis one can more clearly specify the objectives and elements of design and list any outstanding problems.

Where the development of a training system is concerned, the main interest is on specifying and characterizing the important competencies and skills that should be targeted, as well as the prospective population of trainees for whom the system should be designed and developed. Hence although the same methods and task analysis tools that are employed in traditional user-centered task analysis are also employed in training-oriented analysis, there is a major shift of focus and perspective. Training systems are developed to enable learning and to provide alternative preparatory experience for the performance of tasks in the actual environment. In addition, for almost all tasks for which on-the-job training is not possible and preparatory training environments are developed, it is impossible to construct a full-scale complete analogue of the real-life task (think for example of surgery, competitive rowing, or the industrial maintenance job). Hence, by default, training systems constitute partial realizations, the parts of which and the mapping rules from the real-life task and operational environment should be well stated and justified. The driving factors for training task analysis are the skills and competencies that need to be acquired, and the elements to be captured and instantiated in the training platform.

As indicated above, in our study of developing multimodal platforms for perceptual motor skills, nine training platforms for six tasks have been developed. Rowing (ROW), juggling (JUG), maxillofacial surgery (MFS), upper limb rehabilitation (ULR), industrial maintenance and assembly (IMA), programming by demonstration (PBD), and the IMA and PBD demonstrators include both VR and

AR platforms; the ULR demonstrator includes exoskeleton (Exos) and bimanual (BM) systems. The scientific and practical challenges for the training-oriented task analysis of the six candidate tasks have been the explication of a unified framework of skills and subskills that can serve to orient and direct the conduct of task analysis. Of interest was the significant set of sensory motor and cognitive competencies, which are embedded in expert performance of a selected group of tasks. A brief description of the developed framework is presented below.

SKILL-ORIENTED UNIFIED FRAMEWORK

Drawing upon the existing literature and an initial review of the group of tasks, two categories of a conceptual and reference framework have been developed: sensorimotor and cognitive skills. Table 2.1 presents the categories and the included subskills, followed by their brief description. A detailed discussion of the two categories and the interrelations between sensorimotor and cognitive components in the formation of skills is beyond the scope of this chapter. This is presented in Chapter 3.

The two skill categories and the included subskills have guided the analysis of all tasks in the SKILLS project for which training platforms were developed. They served as the building blocks and targets for the design and development of the engineering platforms, as well as the training scenarios and protocols. The link between training platforms and targeted skills is briefly depicted in Table 2.2.

Accelerators and Training Protocols

The term *accelerator* is used to refer to variables that are introduced and implemented to facilitate, assist, and improve learning. The term *training protocol* is employed to describe training schedule, duration, selected tasks scenarios, difficulty manipulations, and their order of presentation. It should be recognized that the development of multimodal VR training platforms gives trainers and engineers multiple options to introduce feedback, guide information, augment, and assist in facilitating and improving training. Identification and application of these options give the

---

TABLE 2.1

Sensorimotor and Cognitive Subskills Included in the Unified Conceptual Framework of the SKILLS Project

Sensorimotor Subskills	Cognitive Subskills
Bimanual coordination	Control flexibility and attention management
Hand-eye coordination	Coping strategies and alternative response schemas
Interpersonal coordination	Memory organization, structure and development of knowledge
Perception-by-touch	schemas
Prospective control	Perceptual observational, procedural skills
Proximal-distal coupling	
Respiratory/movement coupling	
Fine force control	

TABLE 2.2

Description of Six Multimodal Virtual Reality (VR) Training Platforms and Their Designed Accelerators

	Training Focus	Targeted Population	Accelerators
ROW (Chapter 11)	Acquisition of basic rowing skills, effort and energy management, interpersonal coordination	Novice and intermediate rowers	Online visual spatial trajectory of rowing pattern (Fd) Online vibration directive of rowing pattern (Fd) Adjustable auditory pacer of the locomotors/respiratory coupling (Rhythmic Pacer) Visual director of energy expenditure (Fd) Visual and haptic information of interpersonal coordination (Rhythmic Pacer)
JUG (Chapter 12)	Attention management of multiple moving objects, spatial temporal relationship, bimanual rhythmic coordination	Novice jugglers	Tactile-auditory rhythm trainer of juggling coordination (Rhythmic Pacer) Training at slow and gradually increasing task speed (Task processing time) Systematic exploration of the spatial temporal components of the K dwell ratio (Control strategy)
MFS (Chapter 13)	Fine control of force application, use of fine-graded touch and visual information	Trained surgeons	Feedback on forces and torques applied to the tool (Fd) Visual feedback on performance from an "impossible" anatomical point of view (Feedback) Performance feedback relative to optimal performance lines (Feed) Multimodal feedback to enhance sensitivity to compliance and vibration change (Feedback)
IMA (Chapter 15)	Acquisition of procedural skills in virtual environment and via a remote augmented reality training	Technicians and machine operators	Including haptic in 3D VR training (Haptic Enactin) Adding abstract representation to enaction (Cognitive Aid) Introducing direct visual aid (pointer) (Visual Director) Adding images of parts (Cog. Aid) Adding rotational haptic hints (Haptic Enaction) Augmenting enaction by theoretical instructions (Cognitive Aid)

TABLE 2.2 (continued)
Description of Six Multimodal Virtual Reality (VR) Training Platforms and Their Designed Accelerators

	Training Focus	Targeted Population	Accelerators
PBD (Chapter 16)	Exploring and adapting behavior to the motion and compliance constraints of a robotic arm	PBD robot operators	Online indicators of approaching singularity (Feedback) Voluntary exploration of singularity (Control strategy) Haptic exploration of compliance parameters setting (Hp. Enaction)
ULR (Chapter 14)	Using robotic technology and VR to expand therapeutic options/ interaction with patients in rehabilitation	Patients undergoing limb control physiotherapy	Task selection Online continuous feedback (Fd) Motion adaptation (Fd, Motivation)

Notes: ROW refers to rowing, JUG to juggling, MFS to maxillofacial surgery, IMA to industrial maintenance application, PBD to programming by demonstration, and ULR to upper-limb rehabilitation.

trainer the power to facilitate and improve training, in modes and formats that do not exist or are not possible in real-life task performance. In addition, multiple options exist to engineer, order, sequence, and control the protocol of training, its content, sequence, difficulty, and duration. All of these are inherent parts of the design, engineering, conduct of training, and evaluation of each training platform.

SIX CASE STUDIES OF MULTIMODAL VIRTUAL REALITY PLATFORMS

Table 2.2 presents a brief summary of the way in which the general framework, concepts, and principles discussed thus far have been applied to the development of training platforms for six selected tasks. Briefly described for each task are the aspects and competencies that were focused upon, the selection of sensorimotor and cognitive subskills, the identification of the targeted trainee population, and the introduction of accelerators.

ASSESSING ACCELERATORS AND EVALUATING TRAINING

Across the six platforms there have been more than 30 implemented accelerators, 12 of which (the dominant category) are feedback indicators.

Feedback accelerators capitalize on the elaborated measurement and capturing techniques incorporated in each of the demonstrators, to provide trainees with experiential, online feedback on their performance which could not have been presented otherwise. In most cases, the information provided to the learner is enacted by the learner himself or herself (e.g., energy consumption in rowing described in Chapter 11), and the feedback relates to the discrepancy between current and target values. Target values can be derived from analytic or expert models, or reflect objective parameters such as targeted rhythm (e.g., metronome) or errors of path tracking. Feedback indicators may vary in their modality of presentation (visual, tactile, auditory), time mode (continuous versus intermittent), and reference point (trainee performance level, hitting boundaries or constraints, correspondence to expert or optimal performance models). There are three important tests for the value of feedback-based accelerators: (1) the relevance of the information and type of guidance to learning; (2) the ability to improve learning without developing dependence on the feedback presence, which will degrade performance immediately once this feedback is removed; (3) the presence of feedback that should not distract or interfere with the regular modes of performing the trained task. These aspects have been examined in the platforms evaluation studies.

In many cases the study of accelerators and their comparative effects present interesting theoretical conjectures and contrasts. For example, cognitive aids of different types, which have been implemented in the IMA training platform, are natural accelerators for the training of procedural and memory-based skills, the focus of this demonstrator. The addition of haptic information as an accelerator provides an interesting test for the influence of motor enaction on the organization of knowledge in memory and its activation. Recent studies have shown the divergence of semantic-based and action-based representations in memory (e.g., Koriat & Pearlman-Avnion, 2003). The rowing and juggling platforms include an interesting evaluation of the use of rhythmic pacers as accelerators—the Juggling and PBD platforms both test accelerators based on executive control training approaches (Gopher, 2007). These are some of the more general and deeper scientific questions and contrasts that emerge and must be addressed when assessing the accelerators in the different proposed evaluation plans.

Assessing the value of the developed platforms for skill acquisition and the best ways of applying them in training is a multifaceted task. The four basic evaluation aspects that need to be examined are (1) a comparative evaluation of the differential experience of performing the same tasks on the VR platform and in the real world, (2) evaluation of the contribution of accelerators, (3) assessment of training protocols that will maximize learning and skill acquisition on a platform, and (4) transfer of training studies. The first type of evaluation aims to identify the similarities and differences between performing the same tasks (e.g., rowing, juggling, drilling) in the real world and in the VR training platform. Such an assessment is crucial to better understand the differences between acquiring a skill in the virtual and in the real environments, and the possible implications of these differences on the use of the VR platform in training and transfer. The main question is whether expert performers can comfortably employ the same form of behavior and execute their acquired skills in the virtual as in the real environment. This question extends much beyond the subjective feeling of immersion or presence.

CONCLUSION

In summary, it is clear that from a human performance as well as skill acquisition perspective, the new multimodal VR technologies offer new and exciting potential for the development of simulators and the training of complex skills. At the same time, they present nontrivial challenges that should be carefully evaluated and avoided. The description and discussion in Chapters 11 through 16 of the six tasks and nine training platforms that were developed within the SKILLS project underlines and explicates the different aspects addressed in this chapter.

REFERENCES

Adams, J., Gopher, D., and Lintern, G. (1977). Effects of visual and proprioceptive feedback on motor learning. *Journal of Motor Behavior,* 9: 11–22.

Anderson, J. (ed.). (1981). *Cognitive Skills and Their Acquisitions.* Hillsdale, NJ: Erlbaum.

Anderson, J. (1982). Acquisition of cognitive skill. *Psychological Review*, 89(4): 369–406.

Auvray, A., Hartcher-O'Brien, J., Tan, H.Z., and Spence, C. (2008). Tactile and visual distracters induce change blindness for tactile stimuli presented on the fingertips. *Brain Research,* 1213: 111–119.

Blake, M. (1996). The NASA advanced concepts flight simulator—A unique transport aircraft research environment. Paper presented at the AIAA Flight Simulation Technologies Conference, Ames Research Center, Moffett Field, CA.

Bolia, R.S., D'Angelo, W.R., and McKinley, R.L. (1999). Aurally aided visual search in three-dimensional space. *Human Factors*, 41(4): 664–669.

Bresciani, J.-P., Dammeier, F., and Ernst, M.O. (2008). Tri-modal integration of visual, tactile and auditory signals for the perception of sequences of events. *Brain Research Bulletin,* 75: 753–760.

Burr, D., and Alais, D. (2006). Combining visual and auditory information. *Progress in Brain Research,* 155: 243–258.

Cardullo, F.M. (1991). An assessment of the importance of motion cuing based on the relationship between simulated aircraft dynamics and pilot performance: A review of the literature. In *AIAA Flight Simulation Technologies Conference* (pp. 436–447). New York: American Institute of Aeronautics and Astronautics.

Chase, W.G., and Ericsson, K.A. (1981). Skilled memory. In J. Anderson (ed.), *Cognitive Skills and Their Acquisitions,* Chapter 5. Hillsdale, NJ: Erlbaum.

Chopra, V., Gesink, B.J., and de Jong, J. (1994). Does training on an anaesthesia simulator lead to improvement in performance? *British Journal of Anaesthesia* 73(3): 293–297.

Diederich, A., and Colonius, H. (2004). Bimodal and trimodal multisensory enhancement: Effect of stimulus onset and intensity on reaction time. *Perception and Psychophysics,* 66(8): 1388–1404.

Doyle, M.C., and Snowden, R.J. (2001). Identification of visual stimuli is improved by accompanying auditory stimuli: The role of eye movements and sound location. *Perception,* 30(7): 795–810.

Drury, C.G., Paramore, B., Van Cott, H.P., Grey, S.M., and Corlett, E.N. (1987). Task analysis. In G. Salvendy (ed.), *Handbook of Human Factors* (2nd ed.), pp. 370–401. New York: Wiley.

Ernst, M.O., and Bülthoff, H.H. (2004). Merging the senses into a robust percept. *Trends in Cognitive Science,* 8(4): 162–169.

Fort, A., Delpuech, C., Pernier, J., and Giard, M.-H. (2002). Early auditory–visual interactions in human cortex during nonredundant target identification. *Cognitive Brain Research,* 14: 20–30.

Gaba, D.M. (2005). The future vision of simulation in health care. *Journal of Quality and Safety in Health Care,* 13(S1): i2–i10.

Godley, S.T., Triggs, T.J., and Fildes, B.N. (2002). Driving simulator validation for speed research. *Accident Analysis and Prevention,* 34(5): 589–600.

Gonot, A., Chateau, N., and Emerit, M. (2006). Usability of 3D-sound for navigation in a constrained virtual environment. In *Proceedings of the 120th Convention of the Audio Engineering Society.* Paris, France, preprint 6800.

Gopher, D. (2006). Control processes in the formation of task units. In Qicheng Jing (ed.), *Psychological Science around the World,* Volume 2, Social and Applied Issues. Oxford: Oxford Psychology Press. A chapter based on a keynote address given at the 28th International Congress of Psychology.

Gopher, D. (2007). Emphasis change as a training protocol for high demands tasks. In A. Kramer, D. Wiegman, and A. Kirlik (eds.). *Applied Attention: From Theory to Practice.* Oxford: Oxford Psychology Press.

Holden, M.K. (2005). Virtual environments for motor rehabilitation: Review. *CyberPsychology and Behavior,* 8(3): 187–211.

Howell, J.N., Conatser, R.R., Williams, R.L., Burns, J.M., and Eland, D.C. (2008). The virtual haptic back: A simulation for training in palpatory diagnosis. *BMC Medical Education,* 8: 14.

Kaiser, M.K., and Schroeder, J.A. (2003). Flights of fancy: The art and science of flight simulation. In P.M. Tsang and M.A. Vidulich (eds.), *Principles and Practice of Aviation Psychology.* Hillsdale, NJ: Lawrence Erlbaum.

Koriat, A., and Pearlman-Avnion, S. (2003). Memory organization of action events and its relationship to memory performance. *Journal of Experimental Psychology: General,* 132(3): 435–454.

Lalanne, C., and Lorenceau, J. (2004). Crossmodal integration for perception and action. *Journal of Physiology–Paris,* 98: 265–279.

Mantel, B., Bardy, B.G., and Stoffregen, T.A. (2005). Intermodal specification of egocentric distance in a target reaching task. In H. Heft and K.L. Marsh (eds.), *Studies in Perception and Action VIII* (pp. 173–176). Hillsdale, NJ: Erlbaum.

Meyer, D., and Kieras, D. (1997). A computational theory of executive cognitive processes and multiple-task performance: Part I. Basic mechanisms. *Psychological Review,* 104(1): 3–65.

Murray, M.M., Molholm, S., Michel, C.M., Heslenfeld, D.J., Ritter, W., Javitt, D.C., Schroeder, C.E., and Foxe, J.J. (2005). Grabbing your ear: Rapid auditory-somatosensory multisensory interactions in low-level sensory cortices are not constrained by stimulus alignment. *Cerebral Cortex,* 15(7): 963–974.

Oviatt, S. (2002). Multimodal interfaces. In J. Jacko and A. Sears (eds.), *Handbook of Human-Computer Interaction* (pp. 286–304). Hillsdale, NJ: Erlbaum.

Reiner, M. (2004). The role of haptics in immersive telecommunication environments. *IEEE Transactions on Circuits and Systems for Video Technology,* 14(3): 392–401.

Riva, G. (2006). *Virtual Reality. Wiley Encyclopedia of Biomedical Engineering.* New York: Wiley.

Romano, D.M., and Brna, P. (2001). Presence and reflection in training: Support for learning to improve quality decision-making skills under time limitations. *CyberPsychology and Behavior,* 4(2): 265–278.

Salvendy, G. (ed.). (1997). *Handbook of Human Factors* (2nd ed.), Chapters 52 to 55. New York: Wiley.

Sanchez-Vives, M.V., and Slater, M. (2005). From presence to consciousness through Virtual Reality. *Nature Reviews,* 6: 332–339.

Sarter, N.B. (2006). Multimodal information presentation: Design guidance and research challenges. *International Journal of Industrial Ergonomics,* 36: 439–445.

Schmidt, K., and Bjork, D. (1992). New conceptualization of practice: Common principles in three paradigms suggest new concepts for training. *Psychological Science*, 207–217.

Shneiderman, B. (1998). *Designing the User Interface: Strategies for Effective Human-Computer Interaction* (3rd ed.). Reading, MA: Addison-Wesley.

Spence, C., Pavani, F., and Driver, J. (2004). Spatial constraints on visual-tactile cross-modal distractor congruency effects. *Journal of Cognitive, Affective, and Behavioural Neuroscience,* 4(2): 148–169.

Stein, B.E., and Meredith, M.A. (1993). *The Merging of the Senses.* Cambridge, MA: MIT Press.

Stoffregen, T.A., and Bardy, B.G. (2001). On specification and the senses. *Behavioral and Brain Sciences,* 24: 195–261.

Stoffregen, T.A., Bardy, B.G., Smart, L.J., and Pagulayan, R.J. (2003). On the nature and evaluation of fidelity in virtual environments. In L.J. Hettinger and M.W. Haas (eds.), *Virtual and Adaptive Environments: Applications, Implications, and Human Performance Issues* (pp. 111–128). Mahwah, NJ: Erlbaum.

Tindall-Ford, S., Chandler, P., and Sweller. J. (1997). When two sensory modes are better than one. *Journal of Experimental Psychology: Applied*, 3(4): 257–287.

Weller, J.M. (2004). Simulation in undergraduate medical education: Bridging the gap between theory and practice. *Medical Education,* 38: 32–38.

Wichter, S., Haas, M., Canzoneri, S., and Alexander, R. (1997). Keyboarding skills for middle school students. An unpublished manuscript. University of Michigan-Dearborn. Available at www.umd.umich.edu/soe/maaipt/keyboard.html.

Witmer, B.G., and Singer, M.J. (1998). Measuring presence in virtual environments: A presence questionnaire. B.G. Witmer (ed.), *Presence*, MIT Press, Cambridge, MA, pp. 225–240.

Yechiam, E., and Gopher, D. (2003). A strategy-based approach to the acquisition of complex perceptual motor skills. Presented at the 47th Annual Meeting of the Human Factors and Ergonomics Society. Denver, CO, October, 2003.

3 Dynamics of Skill Acquisition in Multimodal Technological Environments

Benoît G. Bardy, Julien Lagarde, and Denis Mottet

CONTENTS

LEARNING SIMULATORS AND TRAINING SCENARIOS

Figure 3.1 illustrates six technological human-machine interfaces developed by a European consortium of researchers in movement and cognitive sciences, robotics, interaction design, as well as engineers and industrials, from 15 institutions,[*] aiming at accelerating and transferring the (re-)learning of complex coordinative skills in virtual and real environments. Such multimodal virtual situations and human-machine training devices are not uncommon in the nascent twenty-first century to simplify, learn, train, maintain, and transfer specific skill elements in domains as various as

---

[*] The European SKILLS integrated project (2006–2011) (www.skills-ip.eu).

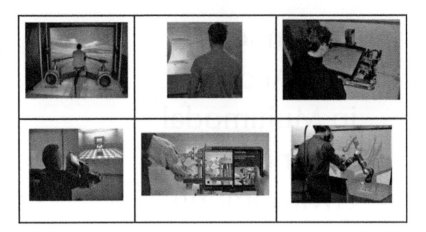

FIGURE 3.1 Six multimodal virtual reality platforms allowing the acquisition and transfer of perceptuo-motor skills in the domains of sports (rowing), entertainment (juggling), surgery (maxillofacial), rehabilitation (upper-limb), industrial maintenance (assembling/disassembling), and programming by demonstration (prototyping).

sports, entertainment, surgery, rehabilitation, or industrial maintenance. Training in technological environments becomes a new era both for trainers and trainees, allowing for the elaboration of new training scenarios and training protocols.

Imagine for instance the two following scenarios that illustrate virtual reality training.

Scenario 1—Learning to Juggle a Three-Ball Cascade

Paula has never juggled before. She stands on the virtual juggling platform in a comfortable position, facing a screen, wearing two specially designed juggling sensors that can record the movements of her hands in six dimensions. Although Paula has no existing juggling skills, she starts moving the hands, enacting three (sonified) balls of different color on the screen. Coordinating progressively the hands themselves, and the hands with the balls, she starts learning the three-ball cascade. Various juggling accelerators are progressively introduced in her training protocol, such as slowing down the virtual balls to perceive the dynamic patterns created by the coordinated motion of the three balls, removing or adding the third dimension during catches and tosses, or injecting at specific moments a cognitive component such as a change of emphasis in what she is led to explore. Online information about current performance or movement execution are added or removed with parsimony using various perceptual channels (haptics, audio, visual).

Scenario 2: Learning to Row as Part of a Team

Mike is a rower with 2 years of experience in team rowing, and he has taken part in several races at both local and national levels. He is well aware that one fundamental component for a successful team rowing is the synchronization among crew members. For fine-tuning this skill component, Mike uses several times a week the SKILLS VR-based rowing training platform. He sits on the sliding seat of the platform, handling the oars, and looking at the visual scene projected on the L-screen in front of him. A team scenario has been prepared in accordance with Mike's level, consisting of three virtual partners rowing in sync and following a predetermined kinematics. Mike starts to row freely, trying to match the imposed kinematics (velocity, stroke rate, etc.) while at the same time rowing in sync with the three avatars, as if he were on a real 4X boat behind the other three athletes. The deviation from synchronized behavior is measured online and feedback is provided to Mike accordingly, at various moments of the training sessions. After a good synchronization index has been reached, the training protocol is adjusted through the implementation of new pacing and distance conditions.

How does learning of these various coordination patterns occur? How do we adequately decompose the skill to be learned in perceptuo-motor primitives that can more easily be trained? What type of information, offline and online, is useful to learn these skill components? Is learning in virtual environments transferable into the real world? Behind the obvious fact that learning requires a large amount of cumulated practice—an effective change in behavior cannot be expected after a limited number of repetitions—a large body of literature suggests that, under specific conditions, VR technology can be an efficient way to improve learning a new skill.

BEHIND THE SCENARIOS

Among many others, these two scenarios are realistic and have been implemented. Details can be found in subsequent chapters of this book. Our main point here is that several fundamental questions are underlying their implementation and are important to answer for successful learning of new coordinative patterns in VR. Four of these fundamental questions are reviewed in this chapter, including (1) the dynamics of skill acquisition in real and virtual environments, (2) the rehabilitation of functional skills, (3) the decomposition of complex perceptuo-motor skills in relevant "trainable" elements, and (4) embodied cognition and enactive learning. In the following sections, we examine these questions in turn.

DYNAMICS OF ACQUIRING A COORDINATION SKILL

Skill is often defined as the capacity acquired by learning to reach a specified goal in a specific task with the maximum of success and a minimum of time, energy, or both. This simple definition, available in any textbook on motor control and learning, suggests that skill cannot be considered a general and abstract ability, but rather a specific and learned capacity operating in a limited ensemble of situations. Many of the skills identifiable in the six examples described in Figure 3.1 require strong *coordination*—between the hands, such as during performing assembly tasks; between the hands and external objects, such as in juggling a three-ball cascade;

between proximal or distal parts of the body, such as when one has to stabilize the postural system in order to operate a robot with the hands in the programming-by-demonstration example; or even between people, such as during team rowing.

The dynamical approach to coordination focuses primarily on the stabilization of a new coordination mode across the multiple elements composing the organism and for an optimal adaptation to the constraints of the situation. The virtually infinite possible coordination modes that an organism can adopt are limited by interacting constraints that can be relative to the environment (e.g., layout and properties of the surrounding objects), to the performer (e.g., biomechanical and physiological properties of the organism), or to the task (e.g., goal of the task, instructions) (cf. Newell, 1985). Coordinated behavior emerges from this coalition of constraints, and an ensemble of principles has been established within the dynamical approach to enhance the acquisition of skill (e.g., Zanone & Kelso, 1992).

SPONTANEOUS COORDINATION COMES FIRST

Facing a novel task, the beginners' behaviors seem attracted by spontaneous coordination modes, determined by either the pre-existing coordination tendencies of the organism (e.g., the tendency of absolute synchronization of the limbs), or by previously acquired coordination. This intrinsic dynamics of the system (Zanone & Kelso, 1992), or *pre-existing movement repertoire* (Faugloire et al., 2006; Sporns and Edelman, 1993), constitutes a backdrop against which the emergence of new patterns will have to occur. The problem of learning is to overcome these spontaneous tendencies to reach other solutions exploiting more efficiently the passive forces of the system (Bernstein, 1967).

ACTIVE EXPLORATION

Learners have to actively explore the workspace of the task in order to discover the properties, regularities, and possible solutions of the problem at hand. During the first trials, the initial (spontaneous) behavior represents an important resource, allowing this active exploration, such as moving the hands in Scenario 1 above. A solution that is often used for teaching new skills is to passively guide the learner by means of direct manipulation or artificial devices such as human-machine interfaces. Passive guidance, however, is a rather inefficient strategy for teaching (e.g., Held & Hein, 1963). Learning requires active involvement of the participant, who has to discover by himself or herself the invariants and regularities related to the task at hand. Learning can be enhanced by "channeling" behavior toward the optimal solution (by virtue of task management), or by manipulating the amount of information to be perceived. But learners need to be able to actively explore the workspace of the task (Newell, 1991).

COMPETITION

Initial coordination tendencies both affect and are affected by the acquisition of a new coordination mode. In this view, learning is not interpreted as the simple addition of the new mode to the pre-existing capabilities but rather as a reorganization of the entire behavioral repertoire. A strong prediction of this approach is that learning-induced modification of the behavioral repertoire can lead to the destabilization of pre-existing skills. Experimental results on the competition principle have been reported in many occasions, for bimanual coordination (e.g., Swinnen et al., 1997; Zanone & Kelso, 1992) and for postural coordination (Faugloire et al., 2006).

Two Learning Routes

Research has demonstrated there are two main ways to adapt to the competition between the intrinsic dynamics and the new pattern prescribed or constrained by the task (Kostrubiec et al., 2006; Zanone & Kelso, 1997). One way is the creation of a new pattern, which after practice is present in the repertoire, cohabiting with the initially stable patterns (the *attractors* of the system). The second way is the shift in one of the initially stable patterns toward the learned pattern. In the first case, referred to as *creation by bifurcation*, if two patterns compose the initial attractor landscape—defined by the number and location of initial stable states—then the new post-learning landscape contains three attractors. In the latter case, called *attractor shift*, the post-learning landscape remains composed of two attractors. One of the two initial attractors has migrated to constitute the new attractor. In the first case, a qualitative change has occurred during learning, while in the second case only a quantitative change has occurred.

REHABILITATION OF SKILL

Rehabilitation in general, and post-stroke rehabilitation in particular, is based on the general principles of motor learning (e.g., Shadmehr & Wise, 2005) and relies on the basic idea that sensorimotor training can harness cortex plasticity (Hallett, 2001). Indeed, after unilateral stroke, the process by which a new skill is learned seems less damaged than the processes by which previously stable motor skills are performed and controlled (Winstein et al., 1999). This probably explains why skill improvement can occur even though physiological recovery has stabilized, and why chronic stroke patients can show considerable motor improvement when involved in novel rehabilitation techniques (Page et al., 2004).

From a general perspective it is rather well accepted, for a damaged or undamaged brain, that motor learning results in long-term cortical reorganization. Sprouting of dendrites with formation of new synapses (Mulder & Hochstenbach, 2001) leads to a functional reconfiguration of brain networks (Bassett et al., 2011). From the perspective of post-stroke rehabilitation, functional recovery achieved by motor learning may be divided into *true* or *compensatory* motor recovery (Krakauer, 2006). *True* motor recovery implies that the synergies at work before the injury (i.e., muscles, joints, segments) are activated. True motor recovery exploits redundancy in the motor networks, either via the remaining undamaged networks or via alternative pathways, and with unmasking of pre-existing cortico-cortical connections. Alternatively and more frequently, *compensation* is the use of alternative synergies to achieve the task in a different manner, most probably relying on fully different brain structures and pathways. It is important to point out that motor learning is required for both types of recovery (Krakauer, 2006), generally following the steps described in the previous section. However, true motor recovery is generally the main goal of post-stroke rehabilitation. In stroke patients, rehabilitation of the paretic limb (arm or leg, for instance) can improve motor function and drive functional plasticity in residual regions of the motor cortex, with the undamaged motor cortex playing an important role in motor recovery (Nudo et al., 1996). However, neuronal plasticity is often experience dependent, time sensitive, and strongly influenced by features of the environment or by the status of the patient, for which motivation and attention can be critical modulators of plasticity (Cramer et al., 2011). Moreover, in many cases maintenance of behavioral gains depends on continued therapeutic exposure.

Motor relearning seems indeed to occur more easily and to be greater when the task is meaningful, and when the training is repetitive and intensive (Hubbard et al., 2009). Intense training favors long-term retention and generalization of the learned motor behavior. Thus, stroke rehabilitation methods should favor intensive and repetitive practice of meaningful tasks. In this context, robotic- and virtual-assisted rehabilitation procedures, which combine intensity, repetitiveness, and online information about crucial components of the task, appear to be efficient (Masiero et al., 2007). Moreover, the added cost of delivering robot/VR therapy is compensated by lower healthcare use costs, even though it remains somewhat uncertain as to what is the exact gain of the cost-effectiveness ratio compared to classic therapy (Wagner et al., 2011). Among others, one distinctive benefit of the approach is its current development toward adaptive upper-limb rehabilitation situations, that automatically modify exercise parameters to account for the specific needs of (different) patients, and implement them to make appropriate decisions about stroke rehabilitation exercises (Kan et al., 2011).

SKILL DECOMPOSITION

The decomposition of complex skills in functional, (re)trainable elements is an important step behind the two scenarios described above, from both theoretical and practical standpoints. From a theoretical perspective, skill decomposition raises the question of the correct level of analysis at which we should capture the elementary yet essential components of skill. Elementary functional units are often described as synergies (Boylls, 1975), coordinative structures (Turvey, 1990), or motor primitives (Mussa-Ivaldi, 1999). Synergies or coordinative structures are temporary (soft) assemblies of elements constrained to behave as a single functional unit (Kugler et al., 1980), while motor primitives are often defined as force fields generated by muscle units (Mussa-Ivaldi, 1999), resulting from the stimulation of hardwire neural circuitry. In line with the dynamics approach to skill acquisition described previously, it has also been suggested that stable attractors (fixed points versus limit cycles) constitute the fundamental building blocks of (continuous versus discrete) movements and skills in humans and other animals or artificial agents (Ijspeert et al., 2007; Schaal et al., 2000). Whether these movement components—soft-assembled synergies, motor primitives, fixed-point and limit-cycle attractors—constitute the right level of analysis for addressing learning issues remains an open question.

NOTION OF SUBSKILLS

Pragmatically, skill decomposition obviously raises the question of which skill elements contributing to the global performance—we call them here the subskills—can be temporarily isolated from their neighbors and receive specific training before being incorporated again into a more general training protocol composed of several interacting subskills and task components. In Table 3.1, we identify 15 skill elements that cross many of the global skills described in Figure 3.1. The first nine subskills (upper section of Table 3.1) are from the sensorimotor repertoire, while the last five subskills (lower section of Table 3.1) are from the cognitive repertoire. In general, sensorimotor subskills are skills that relate to the relationship between perceptual components and motor components, and cognitive skills are related to higher-level cognitive activities that orient, formulate, monitor, and regulate sensorimotor performances. The distinction is partly arbitrary as the two categories are obviously largely interdependent. It is, however, a convenient way to help researchers and designers to design technological tools to enact and learn these skill elements.

Although different perceptual modalities and different types of effectors are involved in performing the skills described in Figure 3.1, the sensorimotor and related cognitive subskills described above are general, abstract, well documented, and cross a large number of domains, from rehabilitation to surgery to industrial maintenance to sports and entertainment. For instance, bimanual coordination is a generic (sensorimotor) subskill that exists in juggling, rowing, and surgery. It includes the concatenation of limit cycle primitives spanning over several muscles and joints. Similarly, procedural subskills are (cognitive) components that exist both in complex surgery and industrial maintenance applications. Among a virtually infinite number of subskills that compose human activities, the 15 elements described in Table 3.1 have been selected because of their key importance for the successful achievement of skilled performance, their possible yet challenging enactment using multimodal virtual reality technology, their coverage of complementary perceptual modalities or effectors, their visible evolution over time and learning, and their anchor in a solid state of the art in basic human movement and cognitive sciences. Each subskill is matter of theoretical and experimental research, is defined by specific variables, and can be captured and rendered using specific hardware and software technologies. As we hope will become evident in the examples below and the subsequent sections of the book, subskills are the main components of efficient and adaptive behaviors in technology-driven learning scenarios. It is through the interaction of these (and other) subskills, under appropriate practice and training conditions, using multimodal interfaces, that skilled behaviors progressively emerge or re-emerge.

LEARNING ROWING IN TEAM USING VIRTUAL REALITY (VR)

The decomposition of a complex perceptuomotor skill into functional subskills now allows focusing virtual reality-based training on specific elements. For instance, the training scenario for team rowing described previously offers the opportunity to test whether following a VR-based training protocol gives better results than a traditional team rowing training. When rowing in a team, the difference in performance

TABLE 3.1
Skill Decomposition into Sensorimotor and Cognitive Components
(With Princeps References)

Sensorimotor Subskills

Balance/postural control (BPC)	The regulation of posture (segments, muscles, joints, etc.) and balance (static, dynamic, etc.) allowing the distal/manual performance to be successfully achieved. BPC is captured by intersegmental and intermuscular coordination, as well as center-of-pressure variables (Nashner & McCollum, 1985).
Bimanual coordination (BC)	The functional synchronization in space and time of the arms/hands/fingers. BC is captured by the relative phase between the coordinated elements and their stability (Kelso, 1995).
Hand-eye coordination (HEC)	The synchronization of eye/gaze/effector with reference to the main information perceptually detected. HEC is assessed by gain, relative phase, and in general coupling variables between eye, gaze, and hand (Biguer et al., 1982).
Interpersonal coordination (IPC)	The coupling between two or more persons. It emerges from a nexus of components including sociality, motor principles, and neuroscience constraints. IPC is assessed by the relative phase between persons (Schmidt & Turvey, 1994).
Perception by touch (PbT)	The coetaneous component of the haptic modality. Various receptors embedded in the skin provide information about mechanical properties (*vibration, compliance,* and *roughness, temperature* and *pain*). PbT is evaluated by psychophysical methods of touch perception (Klatzky & Lederman, 1999).
Prospective control (PC)	The anticipation of future place-of-contact and time-to-contact based on spatiotemporal information contained in optic, acoustic, or haptic energy arrays. PC requires the coupling between movement parameters and information contained in various energy arrays and is measured by time to contact and related variables (Lee, 1980).
Proximo-distal coupling (PDC)	The spatiotemporal coordination of proximal, gross components with distal manipulatory components. PDC refers to the organization of the body underlying arm movements, or to the synergy between arm postures and hand movements. PDC is assessed by cross-relational variables (Berthier et al., 2005).
Respiratory-movement coupling (RMC)	The synchronization of breathing and movement (segments, muscles, joints, etc.) which allows efficient performance. RMC is measured by amplitude, phase, and frequency synchronization patterns (Bramble & Carrier, 1983).
Fine force control (FFC)	The online regulation of the internal forces applied on the surface to successfully reach the goal (drilling, pasting, navigating, etc.). FFC depends on the properties of the surface in relation to the forces developed by the effectors, and is evaluated by the ratio between the two (Faisal et al., 2008).

Cognitive Subskills

Control flexibility and attention management skills (CFAM)	The ability to change response modes and performance strategies, to apply and manage new attention policies in order to cope with task demands and pursue new intentions and goals. CFAM is measured by adjustments to changes in task demand and attention allocation (Gopher et al., 1989).

TABLE 3.1 (continued)
Skill Decomposition into Sensorimotor and Cognitive Components
(With Princeps References)

Coping strategies and response schemas (CSRS)	A vector of importance or attention weights computed over the many subelements of a task, which are associated with the achievement of a specific goal. CSRS is measured by the number and type of strategies to cope with variations in task demand and change in intention (Gopher, 1993).
Memory organization, structure and development of knowledge schemas (MO)	Level of formulated and organized multi-hierarchy, task-specific memory and knowledge bases that facilitate encoding, retrieval, and the conduct of performance. MO is measured by speed and accuracy of encoding; response and decision-making performance; and number, diversity, and speed of generating alternative solutions (Baddeley, 1997).
Perceptual observational (PO)	The ability to detect, sample, and extract task-relevant information from the environment and perceive static patterns and dynamic regularities. PO is measured by speed, efficiency, amount of conscious supervision, and use of higher-level structures and redundancies (Carlson, 1997).
Procedural skills (PS)	Sequences of ordered activities that need to be carried out in the performance of tasks. Performance of every task can be subdivided into a large number of procedures, the competence in the performance of which is developed with training. PS is evaluated by speed and efficiency of performance and type of supervision (un/conscious) (Anderson & Lebiere, 1998).

between two teams often depends on the ability of the athletes to row together in a highly synchronized manner during the race (e.g., Hill, 2002; Ishiko, 1971). For identical movement patterns, the highest speed of the boat is obtained when rowers' movements are synchronized (i.e., when the continuous relative phase between rowers is null). By using the combination of VR rendering techniques and real-time motion capturing, we recently investigated whether it was possible to learn the specific interpersonal coordination subskill with a virtual teammate on an indoor rowing machine, and to transfer the acquired skill to synchronizing with a real teammate (Varlet et al., submitted) (see Chapter 11 for more details).

In our learning protocol, the synchronization was either spontaneous (no feedback other than the presence of the avatar), or increased by an online visual information embodied on the avatar, giving real-time information about the between-bodies coordination (see Figures 3.2B and C). The results were straightforward and indicated that all participants improved their ability to synchronize with a real teammate (pre- and post-test comparison; see Figure 3.2A) following a training protocol with a virtual partner. However, learning was better for the participants who had the embodied feedback available.

RELEARNING POSTURAL COORDINATION AFTER STROKE

Postural rehabilitation of stroke patients is another field in which multimodal technology becomes increasingly efficient. For the recovery of balance and postural stroke-induced deficits, various biofeedback (bioFB) and VR devices have been developed

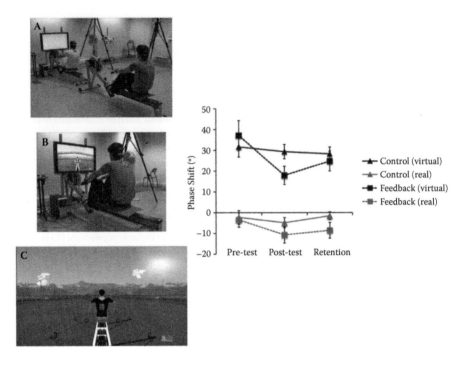

FIGURE 3.2 Left: A VR training protocol for learning team rowing showing real pre-test and post-test (A) training sessions with or without embodied online visual information about the between-rowers relative phase (B,C); right: difference in coordination performance between a control group and feedback group at pre-test, post-test, and retention. (Adapted from Varlet, M., Filippeschi, A., Ben-sadoun, G., Ratto, M., Marin, L., Ruffaldi, E., and Bardy, B.G. Learning team rowing using virtual reality. Manuscript submitted for publication.)

over the past two decades to complete classical therapies (e.g., Dobkin, 2004). Visual and auditory bioFB techniques with inputs from the center of pressure (CoP) have often been to enhance symmetry, steadiness, and dynamic stability (Nichols, 1997). However, the success of these techniques in the long-term recuperation of postural independence is mixed, in part because CoP dynamics is a global factor resulting from the interaction of many endogenous and exogenous variables. Important contributors to COP dynamics are the various postural coordination modes naturally adopted by humans in their interaction with the environment (Horak & Nashner, 1986). In general, the control of standing posture is characterized by rotation of the body around multiple joints, including hips, knee, and ankles, with preferred ankle-hip attractors located around in phase and antiphase (Bardy et al., 1999). Stroke is known to affect the natural ankle-hip coordination landscape (Faugloire et al., 2005) resulting in the complete disappearance of the in-phase pattern, and a loss of stability of the antiphase pattern (Varoqui et al., 2010). Recently, we used a simple postural rehabilitation serious game (see Figure 3.3) to evaluate the dynamics of recovering the lost postural pattern, together with the general stability of the postural system, over a 6-week training protocol (Varoqui et al., 2011). The task required standing

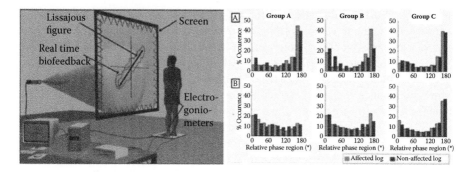

FIGURE 3.3   A postural rehabilitation game consisting of drawing with one's coordinated body visual patterns on a screen (left), and results of a 28-day training protocol illustrating the changes in the corresponding postural landscape (right): Illustrated are the distributions of ankle-hip relative phase values for each leg at pre-test (A) and at post-test (B) for the 0° pattern for the three groups (participants from groups A and B performed the drawing tasks between the two tests with the unaffected and affected leg, respectively; participants from the control group C practiced a stand-up task). (From Varoqui et al. 2011. Effect of coordination biofeedback on (re)learning preferred postural patterns in post-stroke patients. *Motor Control*, 15, 198. With permission.)

participants facing a screen to reproduce with their body an imposed postural coordination mode projected in front of them in an ankle-hip position plane, using a visual coordination bioFB.

In this study, patients were randomly assigned to one of two experimental groups or to a control group. For 28 days, patients from experimental groups followed a training protocol on the two preferred postural patterns using the VR device. One group practiced the drawing game with the unaffected leg while the other group practiced the game with the paretic leg (groups A and B, respectively, in Figure 3.3). The control group, C, practiced a stand-up task for 15 min instead of using the bioFB device. The main results reveal that (1) the two experimental groups rediscovered and improved their in-phase coordination after the relearning compared with the control group, and (2) they showed a related improvement of functional independence measures. Hence, simple treatments using adequate skill accelerators, training protocols, and relying on the adequate subskill—here the ability to synchronize body segments—can be efficient in the rediscovery and maintenance of skill.

EMBODIED COGNITION AND ENACTIVE LEARNING

As mentioned above, the distinction between sensorimotor and cognitive subskills is pragmatically operational but partly arbitrary, as the two categories are largely interdependent. This is because of the natural embodiment of cognitive phenomena into sensorimotor dynamics, and the possibility to enact highly cognitive phenomena using contemporary rendering techniques. Although there are several views on embodied cognition and enaction (Varela et al., 1991; Wilson, 2002), the terms refer to the basic fact that cognition is largely for action, that offline cognition (cognition decoupled from the environment) is largely dependent on the dynamics of body movements and behavior, and that the acquisition of knowledge is largely realized *by doing* and interacting with the environment. Perceptual inputs, for instance (e.g., vision), can elicit covert motor activity in the absence of any task demand. One illustration is judgments of whether a screwdriver is screwing or unscrewing, which have been shown to be faster when the orientation of the handle is consistent with the manual dominance of the observer (de'Sperati and Stucchi, 1997). Other illustrations come from brain imaging studies having shown that the perception of objects affords actions toward these objects with identical activity in cortical regions (e.g., Grezes & Decety, 2002; Grezes et al., 2003). In the same vein, the fact that when individuals observe an action, a brain activation similar to the one arising when actually producing that action is generated in the premotor cortex (Buccino et al., 2001), suggests that embodied cognition plays a role in representing and understanding the behavior of conspecifics (Wilson, 2001), such as in learning by imitation. Hence, perception is not solely a visual or auditory or tactile process, preceding symbolic representations of actions to be performed. What we perceive in the world is influenced not only by, for instance, optical and ocular-motor information, but also by our purposes, physiological state, and emotions. Perception and cognition are thus embodied; they relate body and goals to the opportunities and costs of acting in the environment (Proffit, 2006; Varela et al., 1991). In that respect, theories of embodied cognition and enaction largely rely on the influential work by Gibson in the last 50 years (e.g., Gibson, 1979) calling for a mutual relation between information and movement, between perceiving and acting (e.g., Warren, 1998, for a recent review).

Interesting for the present purpose is that virtual reality technology allows the enactment of new perception-action and cognitive components that are of importance for learning. In the three-ball cascade scenario mentioned above, the spontaneous movements of the hands produced during the very first trial induce technologically controlled multimodal information that can reduce the perceptual complexity of the task, by transforming, for instance, an X-Y-Z movement into a X or an X-Y visual consequence, and so forth. Such *enactive interfaces* therefore represent a means to enhance the conditions for carrying out intuitively manipulative procedures or for learning complex perceptuo-motor skills, to study the conditions for the user of "getting his hands in there and acting" (Stewart et al., 2011), to improve subjective (the feeling of "being there") and objective (*performatory*) fidelity (e.g., Stoffregen et al., 2003), and to evoke new affordances using virtual environments (Gibson, 1977; Stoffregen & Bardy, 2001).

REFERENCES

Anderson, J.R., and Lebiere, C. (1998). *Atomic Components of Thought.* Mahwah, NJ: Erlbaum.
Baddeley, A. (1997). *Human Memory: Theory and Practice.* London: Psychology Press.
Bardy, B.G., Marin, L., Stoffregen, T.A., and Bootsma, R.J. (1999). Postural coordination modes considered as emergent phenomena. *Journal of Experimental Psychology: Human Perception and Performance,* 25, 1284–1301.
Bassett, D.S., Wymbs, N.F., Porter, M.A., Mucha, P.J., Carlson, J.M., and Grafton, S.T. (2011). Dynamic reconfiguration of human brain networks during learning. *PNAS,* 108, 7641–7646.
Bernstein, N. (1967). *The co-ordination and regulation of movement.* London: Pergamon.

Berthier, N.E., Rosenstein, M.T., and Barto, A.G. (2005). Approximate optimal control as a model for motor learning. *Psychological Review,* 112, 329–346.

Biguer, B., Jeannerod, M., and Prablanc, C. (1982). The coordination of eye, head and arm movements during reaching at a single visual target. *Experimental Brain Research,* 46, 301–304.

Boylls, C.C. (1975). A theory of cerebellar function with applications to locomotion. II. The relation of anterior lobe climbing fiber function to locomotion behavior in the cat. COINS technical report number 76-1. Department of Computer and Information Sciences, University of Massachusetts, Amherst.

Bramble, D.M., and Carrier, D.R. (1983). Running and breathing in mammals. *Science,* 219, 251–256.

Buccino, G., Binkofski, F., Fink, G.R., Fadiga, L., Fogassi, L., Gallese, V., Seitz, R.J., Zilles, K., Rizzolatti, G., and Freund, H.J. (2001). Action observation activates premotor and parietal areas in a somatotopic manner: An fMRI study. *European Journal of Neuroscience,* 13, 400–404.

Carlson, R.A. (1997). *Experienced Cognition.* Hillsdale, NJ: Erlbaum.

Cramer, S.C., Sur, M., Dobkin, B.H., O'Brien, C., Sanger, T.D., Trojanowski, J.Q., Rumsey, J.M., Hicks, R., Cameron, J., Chen, D., Chen, W.G., Cohen, L.G., Decharms, C., Duffy, C.J., Eden, G.F., Fetz, E.E., Filart, R., Freund, M., Grant, S.J., Haber, S., Kalivas, P.W., Kolb, B., Kramer, A.F., Lynch, M., Mayberg, H.S., McQuillen, P.S., Nitkin, R., Pascual-Leone, A., Reuter-Lorenz, P., Schiff, N., Sharma, A., Shekim, L., Stryker, M., Sullivan, E.V., and Vinogradov, S. (2011). Harnessing neuroplasticity for clinical applications. *Brain,* 134, 1591–1609.

De'Sperati, C., and Stucchi, N. (1997). Recognising the motion of a graspable object is guided by handedness. *Neuroreport,* 8, 2761–2765.

Dobkin, B. (2004). Strategies for stroke rehabilitation. *The Lancet Neurology,* 3, 528–536.

Faisal, A., Selen, L., and Wolpert, D.N. (2008). Noise in the nervous system. *Nature Reviews Neuroscience,* 9, 292–303.

Faugloire, E., Bardy, B.G., Merhi, O., and Stoffregen, T.A. (2005). Exploring coordination dynamics of the postural system with real-time visual feedback. *Neuroscience Letters,* 374, 136–141.

Faugloire, E., Bardy, B.G., and Stoffregen, T.A. (2006). The dynamics of learning new postural patterns. Influence on pre-existing spontaneous behaviors. *Journal of Motor Behavior,* 38, 299–312.

Gibson, J.J. (1977). The theory of affordances. In R.E. Shaw and J. Bransford (Eds.), *Perceiving, Acting, and Knowing: Toward an Ecological Psychology* (pp. 67–82). Hillsdale, NJ: Erlbaum.

Gibson, J.J. (1979). *The Ecological Approach to Visual Perception.* Boston: Houghton Mifflin.

Gopher, D. (1993). The skill of attention control: Acquisition and execution of attention strategies. In D.E. Meyer and S. Kornblum (Eds.), *Attention and Performance 14: Synergies in Experimental Psychology, Artificial Intelligence, and Cognitive Neuroscience* (pp. 299–322). Cambridge, MA: MIT Press.

Gopher, D., Weil, M., and Siegel, D. (1989). Practice under changing priorities: An approach to training of complex skills. *Acta Psychologica,* 71, 147–179.

Grezes, J., and Decety, J. (2002). Does visual perception of object afford action? Evidence from a neuroimaging study. *Neuropsychologia,* 40, 212–222.

Grezes, J., Tucker, M., Armony, J.L., Ellis, R., and Passingham, R.E. (2003). Objects automatically potentiate action: An fMRI study of implicit processing. *European Journal of Neuroscience,* 17, 2735–2740.

Hallett, M. (2001). Plasticity of the human motor cortex and recovery from stroke. *Brain Research Review,* 36, 169–174.

Held, R., and Hein, A. (1963). Movement-produced stimulation in the development of visually guided behavior. *Journal of Comparative and Physiological Psychology,* 56, 872–876.

Hill, H. (2002). Dynamics of coordination within elite rowing crews: Evidence from force pattern analysis. *Journal of Sports Sciences,* 20, 101–117.

Horak, F.B., and Nashner, L.M. (1986). Central programming of postural movements: Adaptation to altered support-surface configurations. *Journal of Neurophysiology,* 55, 1369–1381.

Hubbard, I.J., Parsons, M.W., Neilson, C., and Carey, L.M. (2009). Task-specific training: Evidence for and translation to clinical practice. *Occupational Therapy International,* 16, 175–189.

Ijspeert, A.J., Crespi, A., Ryczko, D., and Cabelguen, J.-M. (2007). From swimming to walking with a salamander robot driven by a spinal cord model. *Science,* 315, 1416–1420.

Ishiko, T. (1971). Biomechanics of rowing. *Biomechanics II,* 249–252.

Kan, P., Huq, R., Hoey, J., Goetschalckx, R., and Mihailidis, A. (2011). The development of an adaptive upper-limb stroke rehabilitation robotic system. *Journal of Neuroengineering and Rehabilitation,* 8, 33.

Kelso, J.A.S. (1995). *Dynamics Patterns: The Self-Organization of Brain and Behavior.* Cambridge, MA: MIT Press.

Klatzky, R.L., and Lederman, S.J. (1999). The haptic glance: A route to rapid object identification and manipulation. In D. Gopher and A. Koriat (Eds.), *Attention and Performance XVII: Cognitive Regulation of Performance: Interaction of Theory and Application* (pp. 165–196). Cambridge, MA: MIT Press.

Kostrubiec, V., Tallet, J., and Zanone, P.-G. (2006). How a new behavioral pattern is stabilized with learning determines its persistence and flexibility in memory. *Experimental Brain Research*, 170, 238–244.

Krakauer, J. (2006). Motor learning: Its relevance to stroke recovery and neurorehabilitation. *Current Opinion in Neurology,* 19, 84–90.

Kugler, P.N., Kelso, J.A.S., and Turvey, M.T. (1980). On the concept of coordinative structures as dissipative structures. I. Theoretical lines of convergence. In G.E. Stelmach and J. Requin (Eds.), *Tutorial in Motor Behavior* (pp. 3–47). Amsterdam: North-Holland.

Lee, D.N. (1980). The optic flow field: The foundation of vision. *Philosophical Transactions of the Royal Society of London,* B290, 169–179.

Masiero, S., Celia, A., Rosati, G., and Armani, M. (2007). Robotic-assisted rehabilitation of the upper limb after acute stroke. *Archives of Physical Medicine and Rehabilitation,* 88, 142–149.

Mulder, T., and Hochstenbach, J. (2001). Adaptability and flexibility of the human motor system: Implications for neurological rehabilitation. *Neural Plasticity,* 8, 131–140.

Mussa-Ivaldi, F. (1999). Modular features of motor control and learning. *Current Opinion in Neurobiology,* 9, 713–717.

Nashner, L.M., and McCollum, G. (1985). The organization of postural movements: A formal basis and experimental synthesis. *Behavioral and Brain Sciences,* 26, 135–172.

Newell, K.M. (1985). Coordination, control and skill. In D. Goodman, R.B. Wilberg, and I.M. Franks (Eds.), *Differing Perspectives in Motor Learning, Memory, and Control* (pp. 295–317). Amsterdam: North-Holland.

Newell, K.M. (1991). Motor skill acquisition. *Annual Review of Psychology,* 42, 213–237.

Nichols, D. (1997). Balance retraining after stroke using force platform biofeedback. *Physical Therapy,* 77, 553–558.

Nudo, R.J., Wise, B.M., SiFuentes, F., and Milliken, G.W. (1996). Neural substrates for the effects of rehabilitative training on motor recovery after ischemic infarct. *Science,* 272, 1791–1794.

Page, S.J., Gater, D.R., and Bach-Y-Rita, P. (2004). Reconsidering the motor recovery plateau in stroke rehabilitation. *Archives of Physical Medicine and Rehabilitation,* 85, 1377–1381.

Proffit, D.R. (2006). Distance perception. *Current Directions in Psychological Science*, 15, 131–135.

Schaal, S., Kotosaka, S., and Sternad, D. (2000). Nonlinear dynamical systems as movement primitives. *Humanoids2000, First IEEE-RAS International Conference on Humanoid Robots, CD-Proceedings*. Cambridge, MA, Sept. 6–7, pp. 117–124.

Schmidt, R.C., and Turvey, M.T. (1994). Phase-entrainment dynamics of visually coupled rhythmic movements. *Biological Cybernetics*, 70, 369–376.

Shadmehr, R., and Wise, S.P. (2005). *Computational Neurobiology of Reaching and Pointing: A Foundation for Motor Learning*. Cambridge, MA: MIT Press.

Sporns, O., and Edelman, G.M. (1993). Solving Bernstein's problem: A proposal for the development of coordinated movement by selection. *Child Development*, 64, 960–981.

Stewart, J., Gapenne, O., and Di Paolo, E. (Eds.). (2011). *Enaction: Towards a New Paradigm for Cognitive Science*. Cambridge, MA: MIT Press.

Stoffregen, T.A., and Bardy, B.G. (2001). On specification and the senses. *Behavioral and Brain Sciences*, 24, 195–261.

Stoffregen, T.A., Bardy, B.G., Smart, L.J., and Pagulayan, R.J. (2003). On the nature and evaluation of fidelity in virtual environments. In L.J. Hettinger and M.W. Haas (Eds.), *Virtual and Adaptive Environments: Applications, Implications, and Human Performance Issues* (pp. 111–128). Mahwah, NJ: Erlbaum.

Swinnen, S.P., Lee, T.D., Vershueren, S., Serrien, D.J., and Bogaerts, H. (1997). Interlimb coordination: Learning and transfer under different feedback conditions. *Human Movement Science*, 16, 749–785.

Turvey, M.T. (1990). Coordination. *American Psychologist*, 45, 938–953.

Varela, F.J., Thompson, E., and Rosch, E. (1991). *The Embodied Mind: Cognitive Science and Human Experience*. Cambridge, MA: MIT Press.

Varlet, M., Filippeschi, A., Ben-sadoun, G., Ratto, M., Marin, L., Ruffaldi, E., and Bardy, B.G. Learning team rowing using virtual reality. Manuscript submitted for publication.

Varoqui, D., Froger, J., Lagarde, J., Pelissier, J.-Y., and Bardy, B.G. (2010). Changes in preferred postural patterns following stroke during intentional ankle/hip coordination. *Gait and Posture*, 32, 34–38.

Varoqui, D., Froger, J., Pelissier, J.-Y., and Bardy, B.G. (2011). Effect of coordination biofeedback on (re)learning preferred postural patterns in post-stroke patients. *Motor Control*, 15, 187–205.

Wagner, T.H., Lo, A.C., Peduzzi, P., Bravata, D.M., Huang, G.D., Krebs, H.I., Ringer, R.J., Federman, D.G., Richards, L.G., Haselkorn, J.K., Wittenberg, G.F., Volpe, B.T., Bever, C.T., Duncan, P.W., Siroka, A., and Guarino, P.D. (2011). An economic analysis of robot-assisted therapy for long-term upper-limb impairment after stroke. *Stroke*, 42, 2630–2632.

Warren, W.H. (1998). Visually controlled locomotion: 40 years later. *Ecological Psychology*, 10, 177–219.

Wilson, M. (2001). The case for sensorimotor coding in working memory. *Psychonomic Bulletin and Review*, 8, 44–57.

Wilson, M. (2002). Six views of embodied cognition. *Psychonomic Bulletin and Review*, 9, 625–636.

Winstein, C.J., Merians, A.S., and Sullivan, K.J. (1999). Motor learning after unilateral brain damage. *Neuropsychologia*, 37, 975–987.

Zanone, P.G., and Kelso, J.A.S. (1992). Evolution of behavioral attractors with learning: Nonequilibrium phase transitions. *Journal of Experimental Psychology: Human Perception and Performance*, 18, 403–421.

Zanone, P.G., and Kelso, J.A.S. (1997). Coordination dynamics of learning and transfer. Collective and component variables. *Journal of Experimental Psychology: Human Perception and Performance*, 23, 1454–1480.

Section II

---

*Engineering and Technology
of Virtual Reality Training*

4 Perspectives of Multimodal Virtual Reality (VR) Training Platforms

Emilio Sanchez, Jorge Rodriguez, Teresa Gutierrez, Carsten Preusche, and Sara Casado

CONTENTS

INTRODUCTION

Training platforms represent a key part in the overall process of skills acquisition and transfer. They constitute the interface for trainees and therefore play an essential role in the acceptance and success of the approach. All training platforms should provide trainees with clear, useful, and necessary information to facilitate easy interaction between the trainees and the learning tool. It is also desirable that the applied rendering system be multimodal and generate consistent and high-quality output. This chapter focuses on multimodal training systems.

INTRODUCTION TO MULTIMODAL VR TRAINING PLATFORMS

Following the definition of Turk & Robertson (2000), a *multimodal platform* is a system that provides an active human-system interaction (input and output information) through different sensory modalities, specifically visual, haptic, and auditory.

The *multimedia system* is another similar concept. According to Turk & Robertson (2000) a *multimodal interface* supports multiple and simultaneous inputs and outputs (e.g., the use of speech combined with pen-based gestures), while a *multimedia interface* supports only multiple outputs (e.g., text with audio provided to the user).

We define a *multimodal virtual reality platform* as a *multimodal system* that enables the interaction of the user with a synthetic environment modeled in a computer (*virtual reality, VR*).

Multimodal VR platforms are increasingly being used for teaching and training across a wide range of areas including surgery (Howell, 2008), medical rehabilitation (Holden, 2005), air traffic control (Sanderson et al., 2007), and driving (Godley et al., 2002). These multimodal platforms can be used for acquiring new skills or improving existing ones (Derossis et al., 1998; Kneebone, 2003).

If a multimodal VR platform reproduces a real task dedicated to teaching a trainee how to acquire a new skill it is called a *multimodal virtual reality training platform*.

There is no formal categorization of multimodal VR training platforms. As outlined later in this chapter, the platform depends on the type of activity that can be trained with the system, the technology used in the system, the main type of trainee-platform interaction, and the way in which the information is exchanged.

A multimodal VR training platform can make use of different technologies according to the task and user needs. For example, haptic devices can be used to teach or improve skills for drawing and writing (Rodríguez et al., 2008; Younhee et al., 2009), for medical rehabilitation (Alamri et al., 2007), or for training industrial maintenance tasks (Chang et al., 2010).

There is a wide range of benefits from using multimodal VR training systems:

- An *enactive* or embodied *approach* is supported (e.g., Thompson, Varela, & Rosch, 1991), based on *learning by doing* by means of a real-time interaction with the virtual scenario, allowing trainees to undertake the virtual task in an active way. *Learning by doing* is the most natural strategy and is also compatible with the *cone of learning* (also known as the *cone of experience*) proposed by Dale (1969).
- Constraints (mainly of time, cost, and safety) of using the real scenario are eliminated, for example, in industrial maintenance or surgical applications, machines or patients are not in danger during VR training.
- The sense of presence experienced by the trainee is increased. This can help trainees to develop motor and cognitive strategies for following performance in the real world.

- Extra clues (i.e., visual, audio, and haptic aids) are used which are not available in the real world and can facilitate further understanding of some elements of the task.
- Simulation of the task is flexible and can be adapted to the needs of the trainees and to the training goal, for example, removing some constraints of the task in order to emphasize only key aspects.
- Simulations of complications or unexpected events during the execution of the task can be used for the training of alternative response schemes.
- Opportunities are provided for repetition, feedback, and motivation for trainees (Scott, 2005).
- Logging services are used to record the performance of the trainees during the learning process for further evaluation.

However, using multimodal systems for skills training can potentially harm performance in the real world. The greatest potential danger of multimodal systems is that sometimes learners become increasingly dependent on features of these platforms (e.g., the use of extra cues), which may inhibit the ability to perform the task in the absence of VR technologies. In order to solve this problem it is necessary to develop training protocols that control the use of these features in a suitable way.

MULTIMODAL RENDERING REQUIREMENTS

The studies on the effects of behavior of multimodal stimulation and information display can be generally subdivided into the following main categories:

- *Multimodal redundancy and intersensory facilitation:* The use of more than one modality can improve perception and enhance performance. For example, under noisy conditions, observing the speaker's lips movements and gestures can compensate for the lost speech perception (Sumby & Pollack, 1954).

    Reaction times (RTs) for detecting a visual signal are known to be slower in unimodal conditions compared to a synchronized sound-visual condition (Snowden & Doyle, 2001).

    In haptic modality, pairs of auditory and haptic signals delivered simultaneously were detected more quickly than when the same signals were presented unimodally (Murray et al., 2005). In a texture discrimination task, the accuracy of participants was improved (fewer errors) when they received bimodal visual and haptic cues as compared to unimodal conditions in which only a haptic cue or visual cue was presented (Heller, 1982). A trimodal combination of auditory, visual, and haptic stimuli was detected more quickly than the bimodal combination (Colonius & Diederich, 2004).

    To sum up, during the process of integrating and augmenting impoverished sensory cues, information from one sensory channel may be used to augment ambiguous information from another sensory channel (Kim et al., 2001).

- *Dominance of one modality over another*: Vision is the prime and preferred sensory modality for humans (Battig, 1954). However, other modalities can also dominate under certain circumstances. For example, it can be stated that how multimodal information is combined depends on the reliability of the stimulus inputs, and in those cases the more reliable input often dominates (Burr & Alais, 2006).
- *Intermodal information and influence on realism and immersion in virtual environments*: Multimodal interfaces succeed in creating a stronger sense of presence by better mimicking reality, which is multimodal by nature (Brna & Romano, 2001). A single-channel media is relatively sensory-poor and conveys limited and insufficient information to the senses; thus it engenders a lower sense of presence. Conversely, multimodal environments provide a greater extent of sensory information to the observer. This sensorial richness translates into a more complete and coherent experience. And therefore, the sense of being present in a VR is felt stronger (Durlach & Held, 1992; Sheridan, 1992; Witmer & Singer, 1998). The experience is especially felt as real if it includes also haptic (tactile and kinesthetic) sensations (Ho et al., 2000; Reiner, 2004).

    Therefore in *multimodal virtual environments,* the cognitive process of integration induces a multisensory filling in of missing information, in a rather active and creative manner (depending on the user abilities), and this active cognitive integration process results in an enhanced sense of presence.

IMPLICATIONS FOR MULTIMODAL VIRTUAL REALITY TRAINING PLATFORMS

Maintaining multimodal stimulation sources as close as possible to their natural presence in the real environment increases the presence and immersion of performers in a VR environment (Stoffregen et al., 2003). However, the more complex and dynamic a system or an environment, the larger are the difficulties and costs of creating a high-fidelity VR environment. From this perspective, there is a clear distinction between VR environments that constitute alternative work environments to real environments and those developed for the purpose of simulation and training.

In the case of VR as an alternative work environment, the major emphasis is also on the relevance and transfer of experience acquired in the VR environment to the performance in the actual operational environment. Consequently, if fidelity cannot be preserved or is hard to achieve, it is better to avoid it because of the risk of producing illusionary conjunctions and negative transfer (Schroeder & Kaiser, 2003).

In the development of training environments this is the case where it is sometimes desirable for the purpose of training to create deliberate impoverishment and systematic diversions from fidelity (low-fidelity VR environments). This is done to emphasize components, overcome the dominance of one modality over another, and help the operator to develop sensitivities, capabilities, and modes of behavior that are otherwise suppressed in real-life operational conditions. For example, when using the standard computer keyboard, natural dominance of vision causes performers to intuitively adopt a visually guided typing strategy. Acquisition of touch-typing skills based on proprioceptive information feedback from hands and fingers does not develop without deliberate, long, and tedious training (Wichter et al., 1997). Using a secondary task paradigm in which the visually guided typing strategy was made less attractive led to a faster acquisition of touch-typing skills and higher performance levels both at the end of training and in the retention tests (Yechiam et al., 2001).

ARCHITECTURE OF MULTIMODAL VR TRAINING PLATFORMS

A multimodal VR training platform requires the integration of different tools, devices, and the execution of high-performance algorithms for haptic, visual, and audio rendering that have to be well synchronized to provide the trainee a good interaction with the virtual task (Rodriguez & Sanchez, 2008).

With regard to hardware, skill transfer sometimes requires high precision and high-quality haptic interfaces allowing for fine position and force information transfer. In addition to different haptic fixtures, haptic effects have to be implemented; for example, haptic telementoring (online interaction with an expert), playback prerecorded movements/forces, or haptic virtual fixtures created off-line or created in real time by the system.

The aim of visual rendering software modules is to support the large range of hardware setups. The main area of focus needs to be on visual display techniques, augmented reality environments, system scalability, clustering, and often realistic rendering.

Finally at the sound level, a 3D audio engine will increase fidelity.

To sum up, a multimodal VR training platform consists of many independent modules (see Figure 4.1) that need to be coordinated. The agent responsible for this coordination is another module known as middleware.

A *middleware* is a software layer in charge of the communication among the system software modules and provides a hardware abstraction layer to work with several hardware platforms.

A middleware must provide security, identification, authentication, synchronization services, and scheduling facilities at a higher level than those provided by the

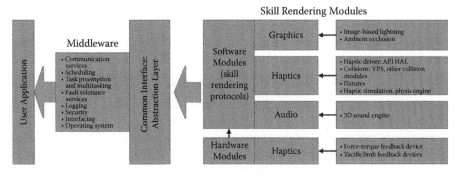

FIGURE 4.1   VR training platform based on a middleware.

operating system. It also must ensure the correct interchange of data between different devices and applications software modules (see Figure 4.1). The middleware should not affect the overall performance of the training system platform and should work in a transparent way.

The use of a middleware facilitates the interoperability. This *interoperability* can be considered from two perspectives:

- *Technical interoperability* will ensure that several systems can be connected together and check that the information is sent properly. But it is not guaranteed that the receiver will understand and interpret the transmitted data correctly.
- *Data interoperability* will support the semantics standardization of the transmitted data among different systems—that is to say, it has to set a common dictionary.

As the training platforms implicitly consider the direct interaction with the trainee, the challenge is focused on getting *real-time performance* of the middleware. Thus, the overall system performance is highly dependent on it. A long list of middleware packages is available, including commercial ones (e.g., VIRTOOLS), middleware from the public domain (e.g., CORBA) or internal developments from technological centers and universities (e.g., AMIRE, AVALON-INSTANT REALITY, CollGate, CORBA, OpenSceneGraph, XVR, etc.).

The emergence of a wide range of tools and devices makes it necessary to provide mechanisms such as special APIs (Application Programming Interfaces) that allow the interoperability, while removing from the application all dependencies on particular configurations and drivers of the tools and devices used. This would facilitate the use and integration of new tools with the application. For example, several haptic APIs have been developed for research applications, such as Chai3D or H3DAPI* that are Open Source and support the most popular haptic devices such as delta.x, omega.x, falcon, phantom, and freedom6. A further example is the HAL (Haptic Abstract Layer) API (Gutiérrez et al., 2010). This API also provides a common development interface that is independent from the drivers of each haptic device and allows the integration of new devices, thus providing greater flexibility to the system. It also manages the use of more than one haptic device simultaneously in the same session in order to simulate bimanual tasks. Examples of supported haptic devices are phantom, grab (Avizzano et al., 2003), or LHIfAM (Savall et al., 2004).

TRAINING STRATEGIES WITH VR TRAINING PLATFORMS

A training program on a VR platform is defined by the virtual scenario corresponding to the task to be learned, the set of training strategies used during the training process, and the measures and methods used to evaluate the progress of trainees. The selection of the training strategies for learning a task will depend on the profile of

---

* CHAI3D.org home page: http://www.chai3d.org.

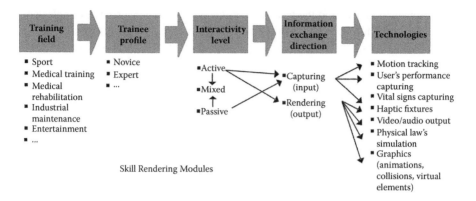

FIGURE 4.2 Taxonomy of VR platform as a training environment.

the trainees and their evolution during the training process and the goal and type of skills involved in the task.

There are four main elements in a VR training strategy: the type of information provided by the strategy to allow trainees to learn the target task, the sensory channel used to transmit this information, the type and degree of interaction from the trainee to perform the virtual task, and finally, the type of communication between trainee and trainer (see Figure 4.2).

TYPE OF INFORMATION PROVIDED BY THE TRAINING STRATEGY

Humans are able to perceive the environment using all their senses. The traditional VR rendering approach aims to model the virtual world as realistically as possible using available technologies resulting in high immersion. Within the context of specific task training, high realism may not be necessary, and the focus for learning shifts to key cognitive and manual ability aspects.

In the training process, the key information for learning is conveyed to the trainee at three different levels: *performance feedback information, guidance information, and enriched information.* The *performance feedback information* approach provides information on the performance of the trainees, giving a better understanding of their actions, in order to correct their errors. Some examples include visual biofeedback and visual feedback. Visual biofeedback on oxygen consumption during a sports activity helps the operative understand efficient gestures to therefore improve efficiency. Visual feedback illustrates the levels of force applied by the trainee on a target, therefore enabling the trainee to comprehensively see any errors committed (in terms of how much larger/smaller the applied force should be). *Guidance information* is used to directly help trainees perform the virtual task. Examples include visual feedback to highlight the target piece to be grasped in an assembly operation (color change of the pieces) and the path that it should follow to avoid damage to the machine (displaying the 3D trajectory), or a haptic guidance system to learn the spatial motion required for drilling through bones or steel levers. Finally, *enriched information* can enhance the reality and provide additional information to help trainees to understand the task and consequence of actions; some examples are visual display of the trajectory that will follow the ball in a juggling task depending on the tossing speed or use of avatars to represent other participants in a team rowing task.

SENSORY CHANNEL USED TO PROVIDE THE INFORMATION IN VR TRAINING

Depending on the available technology, a training strategy can provide information to help the trainees to learn the target task in a *unimodal* way (vision, touch, audio, or even smell in the most advanced systems) or in a *multimodal* way combining different types of stimuli. For example, for the task of learning a 3D trajectory, a VR system could provide information about the target trajectory with visual feedback (displaying the trajectory on the screen) or haptic feedback (applying a haptic fixture to constrain the trainee's motion along the trajectory) or visual-haptic feedback combining the two types of information.

The combined stimuli in a multimodal strategy can provide *redundant* or *complementary* information. In the first approach, the involved sensorial channels provide information about the same item but in a different way to facilitate its understanding and learning, similar to the previous example where both the visual and the haptic feedback provide information about the same feature: the shape of the target trajectory. On the other hand, in the second approach, the involved sensorial channels provide information about different items. For example, for the task of learning a force pattern along a 3D trajectory, the haptic feedback can provide information about the trajectory (haptic fixtures), and the visual feedback can provide information about the level of force to be applied at each point of the trajectory. This approach can also be used for sensory substitution—when one sensorial channel is overloaded another channel can be used to convey additional information.

TYPE OF INTERACTION BETWEEN THE TRAINEE AND THE VIRTUAL SCENARIO: ACTIVE LEARNING VERSUS PASSIVE LEARNING

A training strategy can require different degrees of interaction from the trainees with the virtual scenario to carry out the virtual task: ranging from a totally *passive* to a totally *active* interaction. In the lower limit of this scale, there is *observational learning* (Bandura, 1977), in which trainees do not have any interaction with the virtual scenario and only receive information about how to undertake the task. Following this criterion, *partly observational learning* can be defined as the learning strategy in which trainees do only part of the task and watch how the system does the other part. On the other hand, in the upper limit there are strategies in which the trainees have to perform the virtual task by themselves without receiving any aid from the system. And in the middle, we have strategies where trainees have to perform the virtual task while receiving some help from the system.

The degree of interaction requested in these strategies depends on the criteria used to provide the help (e.g., providing help in a continuous way or just on a trainee's request or based on the user's performance); the type of information provided by the aids; and the method used to provide it. In this way, a strategy that provides *direct information* about the next action (e.g., in an assembly task, changing the color of the target piece that must be grasped) requires a cognitive interaction less active than a strategy that provides *indirect information* about it (e.g., showing an image of the final result of the operation).

The type of trainee interaction to perform a virtual task also depends on the type of simulation implemented in the strategy and its degree of fidelity with the real task. A training strategy can try to replicate the activities of the real task to the greatest degree possible, or remove nonrelevant activities to emphasize the key aspects. It may change the behavior of a relevant variable of the task, for example, slowing the speed of the balls in a virtual juggling task to allow trainees to concentrate on phasing without time constraints.

The type and degree of interaction are important to impede the creation of trainee dependences with the virtual aids that could limit the skills transfer in the real situation where these aids do not exist. The goal of the VR training strategies is to not only correctly perform the virtual task but to allow the transference of the acquired skills to the real task in the real world.

TRAINEE-TRAINER COMMUNICATION

A training strategy can request different types of trainer-trainee communication in order to generate the information that helps the trainee to learn the target task, generally speaking: no communication, off-line communication, or on-line communication (Gutiérrez et al., 2010).

In the *no-communication approach (virtual fixtures)* the information is programmed off-line or generated in real time by the system—some examples are strategies based on the use of animations or virtual fixtures created in-line according to the trainee's actions.

In the *off-line approach (playback)* trainees have access to the information prerecorded by a trainer; an example is the play-back strategy that allows guiding the trainees to reproduce the same motion and actions as the ones conducted by the trainer in a prerecorded session.

Finally, in the *on-line approach (telementoring)* the information is generated in real time according to the information received directly from the trainer. This approach requires the on-line communication between trainer and trainee who can be in the same or in different locations. If the communication is also bidirectional, the strategy allows the trainers to supervise and evaluate the trainees' performance and provide feedback to correct them when necessary. Some examples are the telementoring strategy where information about the trainer's actions is transmitted to trainees to guide them through the task execution, or collaborative strategies where part of the task is done by the trainer and the other part by the trainee and the behavior and actions from the trainee can depend on the trainer's actions (e.g., in shared tasks like playing tennis).

USE OF VR FOR RENDERING SENSORIMOTOR SKILLS

As outlined above, a multimodal system can provide information about the same task through different sensory channels by using the visual, audio, and haptic channels. In this way, each sensory channel can be used to render the same skill. Tables 4.1 and 4.2 illustrate examples of possible mappings among several sensorimotor skills and technologies that can potentially render these abilities.

BOTTLENECKS OF SKILLS RENDERING TECHNOLOGY

After analyzing the mapping of specific skills with the required technology, we arrive at the necessity of implementing specific rendering strategies using the appropriate rendering technology. This section is focused on the main technology bottlenecks that are foreseen. The list of bottlenecks dictates the main issues to be studied and points the way in which the technology assessment experiments should be conducted.

GRAPHICS RENDERING TECHNOLOGY

From the skills analysis we concluded that the principal rendering technologies are visual assistance, virtual marks, and avatars, as they are present in many training platforms that exploit the graphics output.

- In the case of *virtual marks* and *visual assistance,* it is necessary to generate a variety of signals such as arrows to indicate the user's movements. The aim of this visual assistance is to show the trajectory that the user must follow, or to allow the visualization of internal parts with transparencies, or just to depict a mental model of the intended result.
- To accomplish *bimanual coordination* skill when there is an active haptic feedback channel, it is necessary to be able to detect the collision in a scene between the object that is being handled and the virtual hands that represent the mapping of a user's hands in the VR.
- The *level of complexity* for the virtual scenarios is an aspect that needs further discussion. For example, in grabbing a solid object a system needs to calculate when fingers are in contact with the object, the location of the contact points, and the forces applied by the fingers. Let us assume that the task is to move the object from one location to another and that at some point in the trajectory the calculated forces applied by the fingers are too weak to effectively hold the object. What should the system do? Physically, the object should fall and hit the closest surface. According to the materials of the object and the surface it hits, the bouncing may differ in height as well as produce other sensorial outputs such as sound. Can the system be rendered with this level of accuracy? Or, is it better if the system classifies this event as an uncertainty and resets the virtual scenario to a default state? The first approach is closer to reality but requires a more complex rendering than the second approach. Although the second approach is not as realistic as the first, it may be just sufficient from a functional point of view (i.e., all the information critical for the task is still rendered).

TABLE 4.1

Sensorimotor Skills and Rendering Technology Mapping

Skills	Graphics Rendering	Haptic Rendering	Audio Rendering	Combination of Technologies
Balance and posture control	Visual information discrepancy between current and required balance direction (arrows showing how to balance or shift weight) Avatar showing the correct posture to adopt	To display or augment haptically the influence of a good/bad posture, signaling the postural orientation to adopt	Auditory assistance strategies (AAS)	Division of labor: vibro-tactile or audio feedback for orientation and visual avatar for postures
Bimanual coordination	Avatar depicting movements Virtual target markers Visual assistance (difference between the target and the current) Visual guidance of bimanual relative phase	Two haptic devices or two exoskeletons Haptic assistance for hand coordination Collaborative haptic software Use of haptic-props	Auditory displays from the virtual objects handled during assessment of bimanual coordination AAS	Right hand → one haptic device Left hand → data globe/wii combined with an avatar
Hand-eye coordination	Virtual markers	Haptic/vibro-tactile feedback devices Haptic assistance		Decoupling visual representation axis and interaction axis
Interpersonal coordination	Virtual target markers Matching and overlapping of synchronized actions Visual assistance (to show the trajectory of the other person)	Haptic/vibro-tactile feedback devices Collaborative haptic software	Auditory on-line feedback about paddle synchronies	Division of labor between visual information about posture coordination and auditory information about paddle synchronies
Perception-by-touch	Virtual markers Avatars Objects change color to show collisions	Haptic/vibro-tactile feedback devices Haptic assistance	Sound synthesis Realistic audio feedback	Collision detection Haptic device combined with sound synthesis

continued

TABLE 4.1 (continued)
Sensorimotor Skills and Rendering Technology Mapping

Skills	Graphics Rendering	Haptic Rendering	Audio Rendering	Combination of Technologies
Prospective control	Visualization of extrapolated movements (separation of captured and extrapolated movements using different rendering modalities)	Haptic information about time-to-contact	Auditory information about time-to-contact	
Proximal-distal coupling	Rendering of centered and discentered actions on large high-resolution displays (e.g., HEyeWall) Visual information (color, arrows) about proximal movements on distal performance		Presence/absence of auditory information (tone) about optimum coupling between proximal (posture) and distal (arm/hand/fingers) components	Combination of mono/stereo-vision with haptic feedback to evaluate the dominant channel for the exploration of depth information
Respiratory/movement coupling	To show the correlation between captured respiratory and movement information of resulting metadata Visualization of individual energy consumption/energy requirement			
Fine motor control	Visualization of tremor, applied forces, smoothness of motion curve Avatar showing and depicting the correct posture and movements Virtual target markers Visual assistance (transparencies, difference between the target/current)	Haptic/vibro-tactile feedback devices Haptic assistance (to learn with accuracy the trajectory positions)	Realistic audio feedback AAS	Haptic arm device coupled with a grasping force display

TABLE 4.2

Cognitive Skills and Rendering Technology Mapping

Skills	Graphics Rendering	Haptic Rendering	Audio Rendering	Combination of Technologies
Control flexibility and attention management	Virtual target markers	Actuation or switches in the handle to react on events in the task Haptic/vibro-tactile feedback devices	Auditory displays to direct user's attention Realistic audio feedback AAS	Multimodal rendering of attention-distracting events (e.g., overlapping of visual guidance and attention-distracting audio events)
Coping strategies and alternative response schemas	Simulation and rendering of complications and failures	To indicate that multiple choices exist		
Memory organization, structure and development of knowledge schemas	Reduction of visual guidance within different levels of training procedures	Haptic/vibro-tactile feedback devices		
Perceptual-observational	Visual assistance (see internal parts with transparences, show trajectories, etc.)	Haptic devices	Realistic audio feedback	
Procedural skills	Visual assistance (to display the sequence of actions/movements) Virtual target markers Virtual marks (chance of color in objects)	Haptic/vibro-tactile feedback devices Haptic assistance to learn the procedural task (playback sequence, fixtures attractors, etc.)	Realistic audio feedback AAS	Additional actuation or switches to control mode changes in the procedural sequence of the task

Audio Rendering Technology

- *Simulation of complex auditory scenes*: The computational load of spatial sound synthesis scales linearly with the number of sound sources that need to be rendered (Adams & Wakefield, 2007). For each sound source, spatial filters are convolved with the sound representing that source, but rendering a large number of sources can quickly overload a simulation. Also, directly affecting sound rendering may also affect the rendering of other modalities. A realistic rendering of the acoustics in a room is a typical example of rendering several sound sources. This is because reflections from the walls are treated as individual sound sources; each reflection reaches the user from a different direction, and thus each reflection needs to be rendered with a different set of filters.

- *Dynamic aspects*: When virtual spatial sound becomes interactive, dynamic aspects must be carefully considered. For example, in situations where the listener is free to move, the simulated sound field must remain constant relative to the listener's movements. In dealing with the dynamic aspects there are mainly three factors involved: latency (<70 to 80 ms), update rate (>20 Hz), and spatial resolution.

- *Integration with rendering engines for visual and tactile modalities*: Perhaps the major problem in multimodal rendering is the synchronization across the different modalities.

    - *Sensitivity to audiovisual synchronization* is a useful perceptual measure to quantify the available processing time for the generation of virtual auditory displays. Average thresholds for impact sounds, including single and repetitive, have been shown to be about 175 ms. Thresholds for speech presented audiovisually have been shown to be about 200 ms. It seems that requirements for audiovisual synchrony are not more demanding than those obtained from unimodal stimuli such as tolerable latency for sound localization (about 70 ms). In general, sensitivity to audiovisual asynchrony is lower when audio is lagging the visual signal than when audio is leading it. This is beneficial considering that from an ecological point of view the auditory stimulus is always lagging the visual stimulus.

    - For *impact sounds*, the average threshold for detection of audiotactile asynchrony has been shown to be about 24 ms with individual thresholds ranging from 5 to 70 ms. The duration does not seem to have a significant effect, suggesting sound onset as the primary cue for the discrimination of audiotactile asynchrony. For broadband noise, thresholds have been found to be in the range of −31 to 75 ms. Negative values indicate that sound precedes the tactile stimulus. If the haptic stimuli is presented to the whole body, it has been shown that audio delays of 10 to 20 ms are already judged as nonsimultaneous with the haptic stimulus (Martens, 2004).

- *On-line generation of sounds* based on a physical model of interaction that supports the high frequencies of the haptic devices (1 Kz): generating the appropriate sound according to the action done (e.g., suitable sound according to the selected material of two colliding objects).

HAPTIC RENDERING TECHNOLOGY

Kinesthetic haptic rendering technologies suffer from several bottlenecks that need to be addressed as specific research challenges. Considering force feedback devices, there are mainly four types of limitations:

- *Performance limitations of the devices*: To allow natural and intuitive usage, haptic interfaces have to be *transparent* and users must feel as if they were holding the virtual objects directly in their hands. Therefore, a sufficiently *large workspace* is needed with an adequate number of interaction points, including a sufficient amount of force feedback within a wide enough bandwidth, a limited friction and inertia, and a large stiffness.

  Similarly, to deal with tasks that involve the replication of large forces, *high force capacity* is needed, while typically devices feature limitations in the maximum force area.

  This challenge is heightened as they have to be addressed simultaneously.
- *Bimanual capability*: A specific design has to be developed, which is particularly difficult as the hand can partially share the same zones of space. This calls for efficient dimensioning and optimization tools and proper electro-mechanical design.
- *Grasping force feedback*: Another difficult and as of yet unsolved problem is kinesthetic feedback to the human hand. The human hand is a very complex sensor and actor with which a human can manipulate the world in a manifold way. Fortunately often in a given scenario the human does not use the full complexity and dexterity of the hand because of the tool, as this would then require haptic hand interfaces to reflect the required degree-of-freedom with a high precision and force resolution.

  A customized grasping force display with a high-dynamic but compact direct drive motor can be used for rendering the grasping force and augmenting the grasping process with additional training information. The high-dynamic motor can, for example, add a certain vibration to the grasping finger to present virtual fixtures, or forbidden zones in the task-space. The device has to be light-weight, so that it can be easily attached to the handle of a haptic device without disturbing the feedback to the human arm.
- *Simulation limitations*: Most VR software does not allow simulation of complex environments with a sufficient bandwidth for stable haptic rendering. This calls for efficient collision detection and force management algorithms.

  Generally, efficient collision detection and haptic rendering algorithms are needed to deal with massive scenarios (a high number of triangles). In some cases, efficient and realistic haptic rendering algorithms are also requested to work with deformable geometry.

Such simulation bottlenecks are further reinforced when considering applications where several haptic devices have to be used in the same system. In this case, the system has to be able to identify each one of the haptic devices plugged into the system, generate one control at 1 kHz for each device, and calculate collisions among all devices and the VR.

This need of collaborative haptic software to work with several haptic devices simultaneously applies when considering bimanual coordination or interpersonal coordination.

- *Integration limitations*: The integration of different haptic devices has to be done with the minimum software reconfiguration.
- *Ergonomy* of the haptic setups also has to be considered to allow longtime use without strain/pain.
- *Synchronization of haptics* with other modalities has to be made with care to allow an effective and nonconflicting combined multimodal feedback.

Regarding vibro-tactile devices, it can be said that they have the advantage of using small and cheap vibro-motors. Nevertheless, the state-of-the-art systems have three major drawbacks: first, cheap mass production produces a great deal of variability in the quality and motor parameters; second, the small size makes the integration of sensors difficult; and third, every motor requires its own cables, making the system huge and complex.

DELAY IN COMMUNICATION

Connecting a VR system to a human by a multimodal interface means to connect the real analogue world (human) to a discretized digital world (simulation). This connection is made by sensors to detect and measure the human motion, data processing to interpret this data, and calculation of a response to that motion and actors to display the feedback to the subject. All these elements need some detectable time, which results in a delay. This delay is the time that the digital world needs to perceive the real world and to present its feedback.

Each human sensing modality has a different bandwidth, such that the not-perceivable delay depends on the modality (e.g., the human eye perceives a movie with 25 Hz frame rate as fluent). In contrast to that the human skin can detect signals at 10 kHz in its tactile modality. Intended human arm motion has a maximal bandwidth of around 10 Hz (Burdea & Coiffet, 2003), but for force-feedback generally, a sample rate of 1 kHz is seen to be sufficient (Massie & Salisbury, 1994). Given these high requirements, we concentrate on haptics in the following discussion.

The following sources of delay can be found in haptic VRs:

- *Time discretization*: The haptic loop is usually running at 1 kHz, which corresponds to a delay of 0.5 ms.
- *Computation of haptic feedback* (forces/torques): This depends on the algorithm, but one to ten sample periods are typical.

- *Data transmission*: The communication between computers and haptic devices needs real time. It depends on the realization of the system, but several sample periods are typically needed for this data transfer.
- *Inherent effects of haptic device*: The haptic interface adds a lag in the signal during the transformation of data to real action. This delay depends on the design and is caused by, for example, a motor's time constant or structural dynamics.

Summing up, this delay has a destabilizing effect on the haptic interaction between human and VR systems (Gil et al., 2009) that has to be compensated by adding a virtual damping.

EFFECTS OF THE DELAY IN THE INTERACTION

The delay in the system has a high impact on the interaction and therefore in the rendering of accelerators in skill trainers.

First, the above-mentioned additional *damping* leads to a severe *distortion* of the haptic feedback and therefore to a nonrealism of the simulation in the VR system. This means that the training scenario produces sometimes significantly different haptic feedback compared to the real scenario, such that the transfer, especially of motor skills, may not be guaranteed. This aspect has to be accounted for and evaluated in detail.

Second, the *delay between action and feedback* may cause incorrect temporal relationships to be formed. For example, in juggling, the temporal phase between the two hands is of high importance and is determined by the perceived physics of the balls and the desired behavior of the ball sequence (e.g., height of the tosses). A delay in the simulator causes the trainee to form an incorrect relationship between his or her tosses and synchronization of his or her two hands.

Third, the delay in the *reaction of the VR system* to the interaction of the human is the main source of simulator sickness (Draper et al., 2001) that limits the absolute time the trainee can spend with the system.

CONCLUSION

The main conclusion that can be extracted here is the suitability and necessity of multimodal interfaces in virtual training systems because of human nature. However, the present state of the art reveals the number of technological constraints to be high, and it is still far from being able to recreate VR systems with high fidelity.

Many research teams are focused on overcoming these constraints without realizing that the focus should be centered on the development of a good training protocol that re-creates only the realism required at every moment to ensure the progress of the student.

The idea of focusing on the training protocol rather than the technology does not make the research easier, because in the design phase of the multimodal system, we have to combine the discussion results from both technical and human research experts.

REFERENCES

Adams, N., and G. Wakefield. 2007. Efficient binaural display using MIMO state-space systems. In *Acoustics, Speech and Signal Processing, 2007. ICASSP 2007. IEEE International Conference*, Honolulu, HI, 1(I):169-I-172.

Alamri, A., R. Iglesias, M. Eid, A. El Saddik, S. Shirmohammadi, and E. Lemaire. 2007. Haptic exercises for measuring improvement of post-stroke rehabilitation patients. *IEEE International Workshop on Medical Measurement and Applications*, Beneto, Italy, 1–6.

Avizzano, C.A., S. Marcheschi, M. Angerilli, M. Fontana, M. Bergamasco, M.T. Gutiérrez, and M. Mannegeis. 2003. A multi-finger haptic interface for visually impaired people. In *12th IEEE International Workshop on Robot and Human Interactive Communication, Proceedings*, Millbrae, CA, 165–170.

Bandura, A. 1977. *Social Learning Theory*. Prentice-Hall, Englewood Cliffs, NJ.

Battig, W. 1954. The effect of kinesthetic, verbal, and visual cues on the acquisition of a lever-positioning skill. *Journal of Experimental Psychology*, 47:371–380.

Brna, P., and D.M. Romano. 2001. Presence and reflection in training: Support for learning to improve quality decision-making skills under time limitations. *CyberPsychology and Behavior*, 4(2):265–278.

Burdea, G.C., and P. Coiffet. 2003. *Virtual Reality Technology*. Wiley-IEEE Press, New Brunswick, NJ.

Burr, D., and D. Alais. 2006. Combining visual and auditory information. *Progress in Brain Research*, 155:243–258.

Chang, Z., Y. Fang, Y. Zhang, and C. Hu. 2010. A training simulation system for substation equipments maintenance. *International Conference on Machine Vision and Human-Machine Interface (MVHI)*, 572–575.

Colonius, H., and A. Diederich. 2004. Bimodal and trimodal multisensory enhancement: Effects of stimulus onset and intensity on reaction time. *Perception and Psychophysics*, 66(8):1388–1404.

Dale E. 1969. *Audio-Visual Methods in Teaching*. The Dryden Press, New York, NY.

Derossis, A.M., J. Bothwell, H.H. Sigman, and G.M. Fried. 1998. The effect of practice on performance in a laparoscopic simulator. *Surgical Endoscopy*, 12:1117–1120.

Draper, M.H., E.S. Viirre, T.A. Furness, and V.J. Gawron. 2001. Effects of image scale and system time delay on simulator sickness within head-coupled virtual environments. *Human Factors: The Journal of the Human Factors and Ergonomics Society*, 43:129–146.

Durlach, N., and R. Held. 1992. Telepresence. *Presence: Teleoperators and Virtual Environments*, 1(1):109–112.

Gil, J.J., E. Sanchez, T. Hulin, C. Preusche, and G. Hirzinger. 2009. Stability boundary for haptic rendering: Influence of damping and delay. *Journal of Computing and Information Science in Engineering*, 9(1):124–129.

Godley, S.T., T.J. Triggs, and B.N. Fildes. 2002. Driving simulator validation for speed research. *Accident Analysis and Prevention*, 34:589–600.

Gutiérrez, T., J. Rodríguez, Y. Vélaz, S. Casado, A. Suescun, E. Sánchez. 2010. IMA-VR: A multimodal virtual training system for skills transfer in industrial maintenance and assembly tasks. *19th International Symposium in Robotic and Human Interactive Communication (RO-MAN)*, 428–433, Viareggio, Italy.

Heller, M.A. 1982. Visual and tactual texture perception: Intersensory cooperation. *Perception and Psychophysics*, 31(4):339–344.

Ho, C., M.A. Srinivasan, M. Slater, and C. Basdogan. 2000. An experimental study on the role of touch in shared virtual environments. *Transactions on Computer-Human Interaction*, 7(4):443–460.

Holden, M.K. 2005. Virtual environments for motor rehabilitation: Review. *CyberPsychology & Behavior*, 8:187–211.

Howell, J.N., R.R. Conatser, R.L. Williams, J.M. Burns, and D.C. Eland. 2008. The virtual haptic back: A simulation for training in palpatory diagnosis. *BMC Medical Education*.

Kim, J., Y. Choi, and F. Biocca. 2001. Visual touch in virtual environments: An exploratory study of presence, multimodal interfaces, and cross-modal sensory illusions. *Presence: Teleoperators and Virtual Environments*, 10(3):247–266.

Kneebone, R. 2003. Simulation in surgical training: Educational issues and practical implications. *Medical Education*, 37:267–277.

Martens, W.L., and W. Woszczyk. Perceived synchrony in a bimodal display: Optimal intermodal delay for coordinated auditory and haptic reproduction. In *Proc. 10th Int. Conf. on Auditory Display*, July 2004.

Massie, T.H., and J.K. Salisbury. 1994. The phantom haptic interface: A device for probing virtual objects. In *Proceedings of the ASME International Mechanical Engineering Congress and Exhibition*, 295–302, Chicago, IL.

Murray, M.M., S. Molholm, C.M. Michel, et al. 2005. Grabbing your ear: Rapid auditory-somatosensory multisensory interactions in low-level sensory cortices are not constrained by stimulus alignment. *Cerebral Cortex*, 15(7):963–974.

Reiner, M. 2004. The role of haptics in immersive telecommunication environments. *IEEE Transactions on Circuits and Systems for Video Technology*, 14(3):392–401.

Ritter, W., D.C. Javitt, J.J. Foxe, and S. Molholm. 2004. Multisensory visual-auditory object recognition in humans: A high-density electrical mapping study. *Cerebral Cortex*, 14(4):452–465.

Rodríguez, J., C. Vazquez, L. Chirinos, and E. Sanchez. 2008. Haptic system for acquiring drawing skills within a virtual trainer. *Proceedings of the 11th IASTED International Conference CATE*, 440–445, Crete, Greece.

Rodríguez, J., and E. Sánchez. 2008. Technological developments in multimodal interfaces to capture and skills transfer. *Proceedings of IDMME—Virtual Concept 2008*. Beijing, China.

Sanderson, P., M. Mooij, and A. Neal. 2007. Investigating sources of mental workload using a high-fidelity atc simulator. In *Proceedings of the International Symposium on Aviation Psychology* (ISAP2007), 423–468, Dayton, OH.

Savall, J., D. Borro, A. Amundarain, J. Martin, J. Gil, and L. Matey. 2004. LHIFAM—large haptic interface for aeronautics maintainability. In *Video Proceedings of the IEEE International Conference on Robotics and Automation (ICRA)*, Barcelona, Spain.

Schroeder, J.A., and M.K. Kaiser. 2003. Flights of fancy: The art and science of flight simulation. In *Principles and Practice of Aviation Psychology*, 435–471, Erlbaum, Mahwah, NJ.

Scott, O. 2005. Virtual environments for motor rehabilitation: Review. *CyberPsychology and Behavior*, 8:215–216.

Sheridan, T.B. 1992. Musings on telepresence and virtual presence. *Presence: Teleoperators and Virtual Environments*, 1(1):120–125.

Snowden, R. J., and M.C. Doyle. 2001. Identification of visual stimuli is improved by accompanying auditory stimuli: The role of eye movements and sound location. *Perception*, 30(7):795–810.

Stoffregen, T.A., B.G. Bardy, L.J. Smart, and R.J. Pagulayan. 2003. On the nature and evaluation of fidelity in virtual environments. In *Virtual and Adaptative Environments: Applications, Implications, and Human Performance Issues*, eds. L.J. Hettinger and M.W. Haas, 111–128, CRC Press, Boca Raton, FL.

Sumby, W.H., and I. Pollack. 1954. Visual contribution to speech intelligibility in noise. *Journal of the Acoustical Society of America*, 26:212–215.

Thompson, E., E. Rosch, and F.J. Varela. 1991. *The embodied mind: Cognitive science and human experience*. MIT Press, Cambridge, MA.

Turk, M., and G. Robertson. 2000. Perceptual user interfaces. *Communications of the ACM*, 43(3):33–34.

Wichter, S., M. Haas, S. Canzoneri, and R. Alexander. 1997. Keyboarding skills for middle school students. *Unpublished manuscript.*

Witmer, B.G., and M.J. Singer. 1998. Measuring presence in virtual environments: A presence questionnaire. *Presence: Teleoperators and Virtual Environments*, 7(3):225–240.

Yechiam, E., I. Erev, and D. Gopher. 2001. On the potential value and limitation of emphasis change and other exploration enhancing training methods. *Journal of Experimental Psychology: Applied*, 4:277–285.

Younhee, K., Z. Duric, N.L. Gerber, A.R. Palsbo, and S.E. Palsbo. 2009. Demo: Teaching letter writing using a programmable haptic device interface for children with handwriting difficulties. *IEEE Symposium on 3D User Interfaces*, 157.

5 Motion Capture Technologies for Pose Estimation

Paul Smyth and Cristian Canton-Ferrer

CONTENTS

INTRODUCTION

Accurate retrieval of the configuration of an articulated structure from the information provided by multiple cameras is a field that found numerous applications in recent years. Computer graphics technology together with motion capture systems have been extensively used by the cinematographic and video games industry to generate virtual avatars [1]. Medicine has also benefited from these advances in the field of orthopedics, locomotive pathologies assessment, and sports performance improvement [2]. In this field, although markerless motion capture systems have attained accurate and reliable performance in some scenarios [3], only motion capture systems aided by markers placed in landmark positions on the body can produce highly accurate results.

Depending on the type of employed markers, motion capture systems are classified into two groups: nonoptical (inertial, magnetic, and mechanic) or optical (active and passive) systems. Optical systems based on photogrammetric methods are used more than the nonoptical ones, usually requiring special suits embedding rigid skeletal-like structures [4], magnetic [5] or accelerometric devices [6], or

69

multisensor fusion algorithms [7]. Image-based or optical systems allow a relative freedom of movement and are less intrusive. The most usual involve infrared (IR) retro-reflective markers that reflect back light that is generated near the camera's lens [8]. Other optical systems triangulate positions by using active markers that emit a pulse-modulated signal. This allows us to distinguish among markers and automatically label them [9].

Optical motion capture has become a key technology related to human motion analysis and understanding, especially in the fields of gait analysis [2] and gesture recognition [10]. A given skill can usually be associated to a set of motions or gestures, and the manner in which they are performed defines the degree of correctness or point of optimal performance. A sports-related example of this is rowing, where coordination of arms and legs and the temporal evolution of this motion are tightly coupled with the speed of the boat and the efficiency in the use of the rower's energy. Motion capture is a key technology for representing and analyzing these motion patterns, as has been done in the framework of the EU SKILLS project.

There are many users who may benefit from motion capture-based skills analysis. In the context of sports analysis, athletes can assess the degree of performance of their game and pinpoint the weaknesses of their technique. Coaches may then target their training toward improving these specific faults. In medicine, the use of motion capture-based skills analysis is more obvious. Doctors treating patients affected by motor disorders that can be partially or totally overcome using physical rehabilitation can verify the day-by-day evolution of their patients using the proposed techniques.

There are a number of technological challenges that are faced by motion capture systems in the context of skills acquisition, namely the amount of markers to be tracked, the speed of the executed motion, the performance balance between the number of cameras and markers used in the scene, among others. These issues have been tackled through the technique presented in this chapter which makes use of clusters of markers arranged in a preconfigured arrangement, allowing high tracking performance with a low camera count. A summary of the motion capture technologies employed within this project is presented in this chapter, as well as some illustrative examples of its usage.

A data flow diagram is presented in Figure 5.1 that includes the various modules described in this chapter. Essentially, data gathered by motion capture in the form of two-dimensional (2D) coordinates are employed to generate three-dimensional (3D) coordinates through a reconstruction process. After adding temporal consistency to produce 3D trajectories, the pose of several rigid bodies in the scene is estimated in a first stage, and afterward, the whole body pose is computed using subject calibration information. Finally, angles at every joint and possibly consistent 3D anthropometric positions are output to statistically represent a motion-based skill.

MARKER-BASED OPTICAL MOTION CAPTURE

Capturing the motion of a given articulated structure can be achieved by placing a number of distinguishable markers on the body of the performer: In our case, a set of spherical retro-reflective markers are illuminated with IR light and captured by

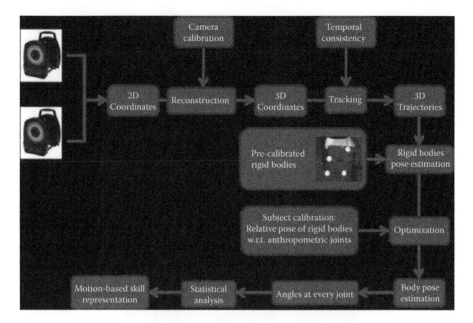

FIGURE 5.1 Data flow of the proposed motion capture SKILLS representation.

IR-sensitive cameras (Figure 5.2). For every camera in the system, a set of 2D detections corresponding to the visible markers in that camera is produced.

Before any motion capture is carried out, a common 3D coordinate system among the cameras must be determined. A calibration process is executed employing a special calibration object known as a wand, whose geometry is known beforehand. This type of calibration process runs a bundle adjustment algorithm that estimates the calibration matrix and lens distortion parameters for every camera similarly, as done in [11]. Finally, using the set of camera calibrations and the 2D detections at every camera, it is possible to obtain the 3D positions of the visible markers in the analyzed volume.

Usually, it is required not only to obtain the 3D positions of every marker but a more semantically coherent description of the observed motion (i.e., when capturing human bodies, it is desirable to retrieve the angles at every joint). To this purpose, a generic human body model is defined, and afterward, a tracking process matches this body model with the incoming data. Because this human body model is generic, it may not faithfully represent the lengths of the bones or the angular span of every joint; hence a body calibration step is usually required employing a *range of motion* (RoM) sequence where the subject exercises all the joints of the body (Figure 5.3).

RIGID BODIES USAGE

Placing a single marker in every body landmark produces 3 degrees of freedom (DoF), namely its x, y, and z coordinates, to be used by the subject calibration and tracking modules down the processing chain. In order to provide a more informative data feed to these modules, it is proposed to replace that single marker measuring

FIGURE 5.2 Elements involved in the motion capture process: IR-sensitive camera with a corona of IR LEDs to illuminate the markers in the scene; typical spherical retro-reflective markers; and the calibration wand.

unit by a *cluster of markers*, also denoted as a *rigid body*, mounted on a small non-deformable platform, thus providing 6 DoF per landmark: x, y, and z coordinates plus its orientation (3×3 rotation matrix). This measuring strategy has already been applied to a number of industrial applications using a few of these rigid bodies [12], and now it is proposed to extend its usage to human motion capture.

FIGURE 5.2   (continued)

FIGURE 5.3   Cluster of markers.

The physical configuration of these rigid bodies is known beforehand, and the placement of the markers on the supporting surface (usually between four and six markers) is unique. In an ideal case, it might be possible to fully distinguish among rigid bodies by looking only at the relative distances among markers. The benefit of this strategy is twofold: on the one hand, more data are provided to the calibration and tracking processes yielding to more accurate results, while on the other hand, it is still possible to estimate the 6 DoF associated with every rigid body even when not all markers are detected (at least only three are required), thus being robust to partial occlusions and able to operate in systems with a low camera count. Moreover, the fact that these clusters are precalibrated and the model associated to them is embedded in the system ensures that they can be tracked straightforwardly. Consistent positioning of these clusters in the same body positions also will allow repeatability of the experiments as well as a quick setup of the whole system (Figure 5.4).

FIGURE 5.4   Upper body motion capture using clusters of markers attached to body landmarks.

The next step is to define a body model that faithfully represents the kinematic structure of the observed subject. Usually this model is based on anthropometric criteria and represents the skeleton as a collection of segments (bones) connected to each other through links (articulations) as shown in Figure 5.5 (top). Finally, every cluster of markers is assigned to a body segment. When tracking a subject using this type of articulated model, the output is usually delivered as the angles at every joint. However, in some applications this level of detail is not required, and the body is modeled as a collection of free joints with no interconnections as shown in Figure 5.5 (bottom). Tracking output in this case is the set of 3D positions of every free joint.

Once the topology and distribution of clusters of markers over the body of the subject under study are defined and encoded into an XML file, it is necessary to calibrate this model toward obtaining the optimal set of parameters (bone lengths and relative position of the clusters of markers w.r.t. the body parts) that will yield an accurate tracking result. This calibration process is carried out by a joint optimization process, and the results are embedded into the XML file describing the model structure and its dynamics. Usually, in order to gather the required data to perform this calibration process, the subject is asked to adopt a standard pose (sometimes, referred as a T-pose) where he or she stands with the arms straight. Afterward, the subject moves every limb (also called RoM), exercising the various joints and thus allowing the system to estimate the angular variability of them. This process also estimates the joint centers or, if not possible, provides an approximate model of them.

This calibration process is very important and should be performed every time a new capture is done. In this way, tracking results will be consistent along time, minimizing the misplacement of some clusters and providing joint angle values that are comparable across different captures.

RIGID BODY TRACKING USING MULTIPLE CAMERAS: ALGORITHMICS

High precision tracking is required to provide meaningful and consistent data to be analyzed at the end of the processing pipeline. The algorithm adopts two parallel

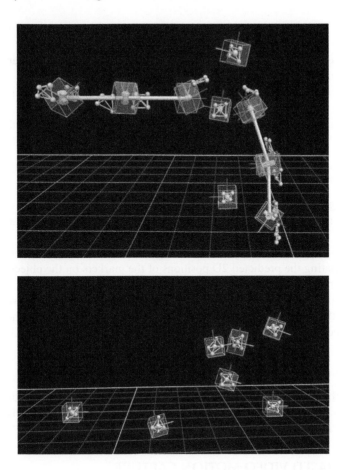

FIGURE 5.5 Two kinematic structures using clusters of markers. On the top, an anthropometric model using joints and segments. On the bottom, a model using only free (6 DoF) joints.

strategies for initiating the object pose estimation process: *boot* and *track*. The pose output from these two algorithms is then used as input to a refinement algorithm which exploits all available 2D observations. Finally, the refined output from the boot and track algorithms are compared, and the best results are reported. In this way, accurate and reliable object tracking can be maintained even with low camera counts, making systems built on this technology very cost-effective in terms of camera hardware.

Boot Algorithm

The boot algorithm is designed to estimate the object pose *without* any a priori information from the previous time-steps. This is typically required when an object has entered the tracking volume, or when it has reappeared after substantial occlusion.

The boot algorithm works as follows:

- Reconstruct 3D points from 2D observations using available camera calibration information.
- Use a rigid subject model to label the 3D points efficiently, exploiting the strong distance statistics between pairs of markers on the model.
- Estimate the approximate pose parameters of the object from labeled points by an SVD method [20].

TRACK ALGORITHM

The track algorithm exploits a priori information from the previous time-steps, potentially allowing it to keep tracking in the face of substantial occlusions. It works by:

- Making a prediction of object pose from previous frames using a motion model
- Calculating the predicted 2D positions of the markers on the object model by using the predicted pose and the camera calibrations
- Matching the expected 2D marker positions to the actual 2D observations
- Directly estimating the approximate object pose parameters from the set of matched 2D observations

REFINEMENT ALGORITHM

This algorithm takes an approximate object pose estimate, a set of 2D observations in each view, and refines the pose estimates and assignments between markers on the model and 2D observations in order to maximize their agreement.

MARKER-BASED VIDEO MOTION CAPTURE

Employing video data for motion capture is a very active field of research where markerless systems have not yet delivered results comparable to the marker-based approaches, in terms of precision and accuracy. While marker-based systems have been widely employed in applications requiring high precision such as orthopedics or cinema, markerless systems have found application in human-computer interaction or gaming [13]. However, there are a number of scenarios where employing markers or clusters of markers is deemed unsuitable, usually when the physical structure of the measuring element may interfere with the required measurement (typically industrial applications, i.e., measuring the flexibility of an airplane wing in a wind tunnel) or when the fixation of the markers may be uncomfortable for the subject under study, for instance, when using a motion capture system on rehabilitation patients. In these cases, it is necessary to provide a less intrusive solution, and a video marker-based motion capture system is proposed (Figure 5.6).

Typically, IR optical motion capture systems make use of retro-reflective spherical markers, while in our video-based system, stickers depicting a dark color circle over a light background will be employed. The steps involved for each camera in our video-based tracking system can be summarized as follows:

FIGURE 5.6 *(See color insert.)* Cluster of markers employed in the proposed video-based motion capture system.

1. *Building a color model for every marker* (if color cameras are available): In order to avoid mismatches between markers, every marker of a given cluster has a specific color, different from its neighbors. Usually, four markers are allocated per cluster, and the assigned colors are those located in the most distant corners of the squared chrominance space, CbCr, with luminance set to Y = 0.5 [14]. YCbCr color space has been selected due to its good decorrelation properties between luma, Y, and chroma, CbCr, allowing us to develop an almost luminance invariant color detector.

2. *Marker detection*: When using a spherical marker in the IR optical motion, its projection onto the image plane is always a circle, thus being easily detected using the Hough transform [15] or the fast approach presented in [16]. However, when employing a flat circular marker, as done in our video-based system, its projection turns out to be an ellipse. Usually, the employed circle/ellipse detection algorithm [16] can detect most of the elliptical shapes present in the scene.

3. *Tracking*: Markers present in the scene are tracked over time, and their state is stored in a vector. When a new measurement is available, the first step is to check whether a direct association between a target in the state vector and any of the measurements can be established, using a shape, distance, and color criterion. For those tracked targets that have no associated measurement, a prediction using the previous states and the context information (distances with other neighbors, statistics of these distances, etc.) is generated.

4. *Creation/destruction of tracks*: In order to have a healthy tracking loop, it is necessary to create new tracks when new clusters appear on the scene and to delete tracks associated with markers or clusters that have no visible markers.

This process will produce a number of timely coherent 2D tracks, but no information about the actual geometric 3D structure of the rigid body has yet been enforced. The next step in the processing chain gathers all these 2D measurements together with the camera calibration matrices and estimates the pose of the rigid body in the scene.

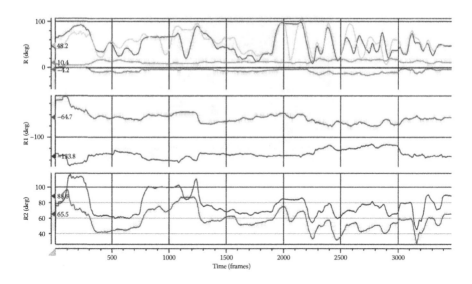

FIGURE 5.7 Time evolution of the joint angles of the arms of a patient executing a rehabilitation task obtained using a motion capture system.

APPLICATIONS WITHIN THE FRAMEWORK OF EU SKILLS PROJECT

Motion capture technologies described in this chapter have been widely employed in a number of skill transfer tasks, namely physical rehabilitation [17], sports analysis [18], or music player performance assessment [19]. As already mentioned, digitally capturing and representing a human skill can allow for assessing the degree of performance attained at performing the associated motion patterns related with the skill under study. There are a number of scenarios where motion capture technologies can be useful, and the EU SKILLS project has addressed two main applications: physical rehabilitation of a patient with motor disorders and sports performance improvement (Figure 5.7).

Upper Limb Rehabilitation

People who have suffered brain lesions resulting in impaired limb coordination usually undertake a rehabilitation therapy. This process usually involves executing several exercises during a number of sessions; quantitatively assessing the improvement in the coordination of the patient from session to session allows the doctor to estimate the recuperation rate and to customize the therapy. In this type of analysis, doctors are generally interested in examining the time evolution joint angles of arms and comparing them across sessions and also comparing them to the patterns of a healthy subject (see Figure 5.6). A number of statistical measures exist to compare two consecutive trials of a patient and assess the degree of improvement and reassure that the therapy is progressing adequately.

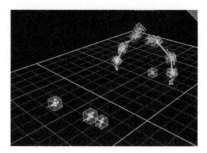

FIGURE 5.8 *(See color insert.)* Motion capture scenario with clusters of markers attached to both the rowing machine and the athlete.

ROWING

Sports performance assessment has greatly benefited from motion capture technologies toward guiding athletes to improve their skills. In this scenario, clusters of markers have been attached to both the rower and the machine, thus allowing us to obtain not only the pose of the rower but also the rower's relative position w.r.t. the machine. Both the 3D position of clusters and the angles at the joints of the subject have been employed to quantify the performance of the rower. Motion capture technology becomes particularly useful when analyzing systems involving multiple rowers where coordination among them is a crucial factor to obtain optimal performance.

Two scenarios have been devised within the row task: the first one is designed to improve the performance of athletes and features between 8 and 12 cameras, and the second is intended for amateur rowers and involves only 2 to 3 cameras (see Figure 5.8).

CONCLUSIONS

Marker-based motion capture technologies can provide useful cues to understand human gestures and ultimately address computer-aided skills transfer and improvement. Optical and video-based motion capture systems employed by VICON within the SKILLS project have been presented, and a novel processing methodology based on the use of clusters of markers has been presented as a way to increase the robustness and accuracy of the tracking while also allowing use of configurations featuring a lower camera count. Two main scenarios have benefited from motion capture technologies: upper limb rehabilitation and rowing.

REFERENCES

1. Baran, I., and Popovic, J. *Automatic rigging and animation of 3D characters.* In *Proc. ACM Int. Conf. on Computer Graphics and Interactive Techniques,* 2007.
2. Cerveri, P., Pedotti, A., and Ferrigno, G. Robust recovery of human motion from video using Kalman filters and virtual humans. *Hum. Mvmt. Sci.,* vol. 22:3, pp. 377–404, 2003.

3. Sigal, L., Balan, A.O., and Black, M.J. HumanEva: Synchronized video and motion capture dataset and baseline algorithm for evaluation of articulated human motion. *Int. J. Comput. Vis.*, vol. 87:1–2, pp. 4–27, 2010.

4. Kirk, A.G., O'Brien, J.F., and Forsyth, D.A. Skeletal parameter estimation from optical motion capture data. In *Proc. IEEE Conf. on Computer Vision and Pattern Recognition*, pp. 782–788, 2005, San Diego, CA.

5. Ascension. http://www.ascension-tech.com

6. Xsens-inertial motion capture. http://www.xsens.com

7. Roetenberg, D. *Inertial and magnetic sensing of human motion.* PhD dissertation, University of Twente, the Netherlands, 2006.

8. Vicon. http://www.vicon.com

9. Raskar, R., Nii, H., Dedecker, B. et al. Prakash: Lighting aware motion capture using photosensing markers and multiplexed illuminators. *ACM Trans. on Graphics*, vol. 26:3, 2007.

10. Kapur, A., Virji-Babul, N., Tzanetakis, G., and Driessen, P.F. Gesture-based affective computing on motion capture data. *Affective Computing and Intelligent Interaction, Lecture Notes on Computer Science*, vol. 3784:pp. 1–7, 2005.

11. Zhang, Z. *A flexible new technique for camera calibration.* Technical report, Microsoft Research, 2002.

12. Tracker. Vicon Motion Capture Ltd., www.vicon.com

13. Shotton, J., Fitzgibbon, A., Cook, M., Sharp, T., Finocchio, M., Moore, R., Kipman, A., and Blake, A. Real-time human pose recognition in parts from a single depth image. In *Proc. IEEE Int. Conf. on Computer Vision and Pattern Recognition*, 2011.

14. Poynton, C. *Digital Video and HDTV.* Chapter 24, pp. 291–292, Morgan Kaufmann, 2003.

15. Duda, R.O., Hart, P.E. Use of the Hough transformation to detect lines and curves in pictures. *Comm. ACM*, Vol. 15:pp. 11–15, 1972.

16. Nel, E., and Stoddart, A. *A blob detection algorithm for video motion capture.* U.S. and European patent, filed in 2008, published in 2009, US 20090180663, EU EP2079054.

17. Hingtgen, B.A., McGuire, J.R., Wang, M., and Harris, G.F. Design and validation of an upper extremity kinematic model for application in stroke rehabilitation. In *Proc. Int. Conf. of the IEEE Engineering in Medicine and Biology Society*, vol. 2:pp. 1682–1685, 2003.

18. Bideau, B., Kulpa, R., Vignais, N., Brault, S., Multon, F., and Craig, C. Using virtual reality to analyze sports performance. *IEEE Comput. Graphics Appl.*, vol. 30:pp. 14–21, 2010.

19. Palmer, C. The nature of memory for music performance skills. In E. Altenmüller, M. Wiesendanger, and J. Kesselring (Eds.), *Music, Motor Control and the Brain*, pp. 39–53, 2006. Oxford, UK: Oxford University Press.

20. Arun, K.S., Huang, T.S., and Blostein, S. Least-squares fitting of two 3-D point sets, *D IEEE Pattern Analysis and Machine Intelligence* September 1987.

6 Skill Capturing and Augmented Reality for Training

Uli Bockholt, Sabine Webel, Timo Engelke, Manuel Olbrich, and Harald Wüst

CONTENTS

INTRODUCTION

Actual developments in information and communication technologies (ICTs) are driven by two trends fostering fundamental paradigm changes in interaction and information processing. On the one hand, smartphones and mobile computing technologies offer highly sophisticated platforms integrating multimodal sensory and powerful central processing units (CPUs); on the other hand, information is structured, geo-referenced linking information to specific locations within our environment. Thus augmented reality (AR) has become a key technology as it analyzes the sensor data (camera, global positioning system [GPS], inertial) to derive the detailed pose of the mobile system, with the aim to correlate our real environment to the geo-referenced information space. Previous research has shown that AR is a powerful technology to support training because instructions on how to perform a specific procedure can be directly linked to the camera-captured environment. Various approaches exist in which the trainee is guided step by step through procedures, but these systems act more as guiding systems than as training systems and focus only on the trainee's sensorimotor capabilities. All these facts lead to the need for efficient training systems that accelerate the learning and acquisition of the skill process. Furthermore, these systems should improve the adjustment of the training process to new training scenarios and enable the reuse of existing training material that has proven its worth. In this chapter a novel concept and platform for multimodal

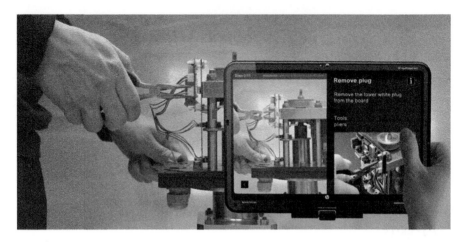

FIGURE 6.1 *(See color insert.)* Augmented reality training of an assembly task.

AR-based training is presented, that fuses capturing and rendering technologies in the following way (cf., Figure 6.1):

- *Capturing*: The camera is used to capture activities (of an expert) within a large-scaled environment. Video cameras are used to track tools or to register motion, and applied forces and torques are registered. The captured trajectories of tools can be represented within a three-dimensional (3D) animation, or the captured information is linked to the real environment using the *Virtual Post-It* metaphor. Virtual Post-Its link contextual information to specific parts of the real world. Starting with forces/torques to be applied and resulting in multimedia illustrations including audio/video files, the Virtual Post-Its can document complex workflows.

- *Rendering*: Within the training scenario the captured information can be presented in two different ways (cf., Figure 6.2): either a 3D-animation illustrating the current work step is directly superimposed on the captured video image (*direct visual aid*), or the tracked Virtual Post-It icon indicates that contextual information is available for a specific part to be displayed if the icon is selected by the user (*indirect visual aid*).

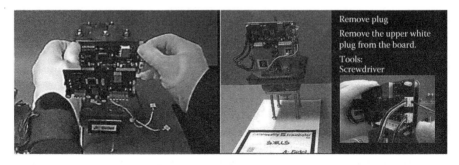

FIGURE 6.2 Direct superimposition of 3D animations (left) or *Virtual Post-It* metaphor.

The AR training system has been designed in a way that content and application flow of the training session are specified within a declarative file based on the X3D ISO standard. This file is loaded and interpreted by the AR system. In this way a new training protocol can be easily specified by the trainer without the need of new programming efforts.

CAPTURING TECHNOLOGIES

Within the SKILLS project, multimodal capturing technologies are used to analyze and qualify skills in different application domains. The capturing technologies are used for two different purposes: activities of experts (e.g., surgeons, sportsmen, or services technicians) are registered to be processed in the digital representation of skill, and capturing technologies are used for the realization of interaction techniques within VR/AR-based training simulators. The capturing technologies include the following aspects:

- *Camera pose tracking*: In computer-vision-based tracking, inside-out as well as outside-in methods are used. In inside-out tracking the camera's position and orientation are calculated relative to detected features within the captured environment. One approach is to place artificial markers, *fiducials*, with special geometric or color properties in the scene. A popular library for tracking using passive fiducials is the ARToolkit (Kato and Billinghurst, 1999). A great deal of research has also been done in the field of markerless camera tracking (Klein and Murray, 2009; Wuest et al., 2005). In outside-in tracking the imaging sensor is mounted outside the space to track and observes the tracking area (e.g., Ferrin, 1991). Especially when multiple cameras are used, outside-in tracking can deliver very accurate results concerning the position, but inside-out is very flexible and not restricted to a specific workspace.
- *Hand tracking*: The tracking of the user's hands is an interesting aspect in the context of manual tasks. One approach for hand tracking uses two cameras in front of a large display screen while the user is performing gestures (Malerczyk, 2008). Another example is the work of Sudderth et al. (2004), in which probabilistic methods for visual tracking of a 3D geometric hand model from monocular image sequences are described. Petersen and Stricker (2009) present a method for real-time hand detection and tracking by focusing on posture invariant local constraints, which exist on finger appearances, instead of considering the hand as a whole. This approach is based on selecting local regions that comply with a number of geometric and photometric posture invariants.
- *Motion capturing*: Motion capturing refers to the digital recording of the movements. Mostly, markers are applied to dynamic subjects, for example human faces, bodies, or robot arms, in order to track their motions (e.g., Kurihara et al., 2002). Tanie and colleagues (2005) enhance the traditional approach by using a suit covered with a retro-reflective mesh instead of using marker spheres. Yoon et al. (2009) present a system for markerless real-time

3D motion capturing to facilitate the analysis of three-dimensional human body movements during a sprint start. Unzueta (Unzueta et al., 2008) presents a method to reconstruct a human's full-body movements based on the positions of the human's end-effectors in order to make it usable within low-cost motion capturing. The first commercially available markerless motion capture system, called OM Stage, is presented by Organic Motion (2011). In addition to optical capturing technologies, nonoptical methods can be used for capturing human movements (e.g., miniature inertial markers can be attached to the human's body, which transmit motion data wirelessly to a computer with which the motion can be recorded, analyzed, or viewed) (e.g., Miller et al., 2004). With the number of attached inertial sensors the quality of the movement that can be reconstructed increases. Inertial motion capture systems provide real-time capturing of full six-dimensional (6D) body motion of a human in large capture volumes but deliver a much lower positional accuracy compared to the accuracy that can be achieved by optical systems.

- *Force capturing*: Hoshi et al. (2006) developed a kind of artificial skin for normal force sensing. It consists of a flexible film containing a network of pressure sensors. The film, which can cover the hand, acts as an artificial skin. The pressure is measured by capacitive effect. The artificial skin also contains a network of capacitive sensors, where each sensor consists of two soft film capacitors lying upon another. Both capacitors vary differently according to the applied stress. This characteristic allows for a simultaneous measuring of the applied normal force and of the properties of the surface over which the force is applied. A further example is given by Yamada et al. who developed an artificial elastic robot finger skin for controlling grasp forces (grip and lift) when weight and frictional coefficient of the grasped object are unknown. Ridges at the surface of the artificial finger skin imitate the ridges of a human finger. Thus the "stick" and "slip" effect that occurs when human fingers grasp an object can be emulated.

Based on this state of the art, the focus of SKILLS is to provide appropriate feedback to the user and to respond to his or her actions during training. Various kinds of information need to be captured addressing those factors that comprise human body motions, applied forces, objects used for interaction, cognitive factors, and much more. The number of aspects involved in the user's performance of a complex skill is enormous, and it is impossible to consider all of them in the training system. Hence the aspects that are most relevant for providing adequate feedback during training have to be chosen and captured. However, the training system should be designed to allow for the integration of many different capturing technologies and the handling of the corresponding captured data, in order to be adaptable to different kinds of skill training (e.g., training of maintenance, rowing, or surgery skills). Thus the capturing technologies can be used for tracking the pose of specific objects, to register activities to be trained, or to realize new interaction paradigms within a training simulator:

- *Registration of activities*: The capturing technologies are used to feed the *digital representation of skill*. This digital representation offers technologies to store and analyzes the nature of a skill. Thus it helps to specify criteria that qualify correctness. In this context, capturing technologies have been developed addressing different modalities that are used to transfer expert actions into the digital repository.
- *Realization of new interaction paradigms*: The capturing forms the basis technology for the realization of interaction within the VR/AR-training systems, and there with an important base for automatic and fluent generation of content. In AR, camera-based interaction based on 6D pose tracking is used, because in mobile applications keyboard/mouse interaction is cumbersome. Using the camera the captured object can be recognized and selected. Also, tracked tools can be used as interaction devices within the VR/AR environment.

CAPTURING OF THE CAMERA POSE

One main challenge in AR is the determination of the camera's pose in order to be able to correctly augment the video image. Thereby, the capturing is differentiated in initialization and tracking. Initialization tries to estimate the pose from just minimal knowledge of the scene (*wide baseline matching*). The tracking then progresses from frame-to-frame observation of points of interest acquired through the initialization, within the video source image, taking the sequential information into account. Types of tracking techniques are then combined. For initialization, edge-based methods are used along with an on-line and off-line reconstruction based on Kanade-Lucas-Tomasi (KLT) features (Wientapper et al., 2011). The computer vision-based tracking is fused with inertial sensors with the aim to enhance stability (Bleser & Stricker, 2009).

CAPTURING OF ACTIVITIES

Activity capturing is a key technology for skill transfer. It goes from simple recognition of movements to reconstruction of complex tasks. Therefore it integrates capturing of instruments and tools, tracking of hands, and recognition of parts and construction components.

- *Capturing of tools*: Many procedures are executed with different tools. Therefore, the tool used within a procedure has to be tracked and identified. To support this, a color-based tracking and recognition algorithm is used. For this algorithm the tool is equipped with colored stripes (cf., Figure 6.3). A color separation algorithm identifies regions of the strips of color within the image. It is able to reconstruct the position and orientation of color-marked tools in real time.
- *Tracking of hands*: In addition to tools, the operator uses his or her hands for performing activities such as maintenance operations on a machine. The developed markerless tracking technologies are utilized to register the operations done with the hand. The introduced color segmentation algorithm can

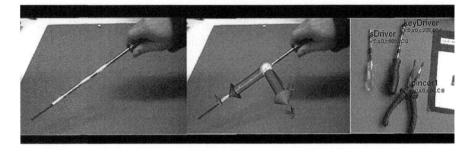

FIGURE 6.3   Color-based tracking of tools (left) and part recognition (right).

also be used as a base for simple hand and activity recognition. Therefore, contours of the segmented images are extracted and analyzed upon previous on-line-generated models. The user has to perform training before the first use in order to generate a user-specific hand recognition model.

- *Part recognition*: Due to the complexity of part identification within the workflow, the task of object recognition is separated from the tracking, and a machine leaning algorithm that is running in parallel to the tool tracking is implemented. The realized approach thereby is a combination of on-line and off-line learning: The 3D-models of the parts will be rendered photo-realistic offline from different points of view and lighting conditions to be correlated to the captured camera images (cf., Figure 6.3 right).

- *Force capturing*: Not only the capturing of motion is relevant for activity capturing but also the registration of applied forces has to be taken into account. In this context, a specific device for torque and force measurement is used (Tripicchio et al., 2010). It supports direct wireless transmission of measurement data into the training protocol. Thus the trainee can learn to estimate correct forces and develop a "feeling" for material. Furthermore, direct feedback can be given to the trainee using graphical user interface (GUI) elements, for instance in the form of quantitative values. A qualitative feedback can be rendered using vibration stimuli provided, for example, through a haptic bracelet (cf., Figure 6.4).

FIGURE 6.4   Force-torque capturing device developed by PERCRO (Pisa, Italy) (right) and haptic bracelet for rendering of tactile feedback developed by German Aerospace Center, DLR (left).

FIGURE 6.5 *(See color insert.)* Distributed AR training session with mobile trainee and remote trainer.

AUGMENTED REALITY AND TELECONSULTATION

This approach for skills transfer using AR offers high impact, as it can be integrated into teleconsultation scenarios in the following way (cf., Figure 6.5):

- The trainee uses the mobile AR equipment to inspect the training environment. In this process, he or she is guided step by step through the training procedure. In the case that a problem occurs during the training session and the trainee does not know how to proceed, the trainee now connects to the remote trainer via network.
- The video images captured with the mobile system of the trainee are now transferred in real time to the trainer. Thus the trainer is able to reproduce the problem of the trainee and is able to modify or enhance the training protocol. For this purpose he or she adds new Virtual Post-Its into the training protocol which can integrate scribbled annotations to the captured video images. The trainer can also link additional multimedia illustrations to the real environment. In this way the training protocol is enhanced and the captured Virtual Post-Its are transferred into the training protocol.

AUGMENTED REALITY-BASED TRAINING

In AR-based training, multimodal elements can be applied for different purposes. For example, to support the trainee in training the fine control of force, it can be beneficial to provide information about the force and torque he or she has to apply and the force and torque he or she is currently applying. The tracking and recognition of hands and tools capture whether the trainee uses the right tools and whether he or she performs the right actions, and if not, error feedback is generated. The technologies have been integrated into a multimodal AR-based training platform. This platform consists of three main building blocks. The multimodal capturing controller handles the data delivered by different capturing systems (e.g., motion capture, camera tracking and object recognition, sound capturing). The interaction processing and application module holds the application and workflow logic and processes the user input. The workflow block inside this module handles the whole workflow of the training task. Furthermore, it embraces concepts for recognizing and interpreting patterns in the capturing data and for refining existing workflows based on captured data. The multimodal rendering controller controls the rendering of the visual, haptic, and audio data.

Thus the following accelerators have been realized, which materialize the training strategies:

- *Virtual Post-Its with information on request*: Instructions during training should be displayed using Virtual Post-Its with a permanently visible Virtual Post-It pointer and a Virtual Post-It content providing additional information on user demand.
- *Context visualization*: The context visualization providing information about the current subtask (i.e., not only about the current step) and about the subgoal to reach should be available in order to enhance the user's mental model building process. The information should contain the condition of the device before and after the subtask.
- *Structure and progress information*: Information about the structure of the training task and the user's progress in the task should be provided. This also enhances the user's mental model building process.
- *Haptic (vibro-tactile) hints*: Additional haptic (vibro-tactile) hints will be presented to the user during training. These hints can be used to clarify information that is difficult to visualize with the available training material (e.g., rotating directions), to present subliminal information, or to provide error feedback (e.g., vibration hints when the user grasps a wrong tool).

SUMMARY AND CONCLUSIONS

The advantage of using AR for training is that the trainee can interact with the real-world objects and simultaneously access the virtual information for guidance (cf., Figure 6.6). Hence the trainee can easily accomplish the mapping between the training and the real task. Furthermore, he or she can perform the actual task

FIGURE 6.6   *(See color insert.)* Augmented reality training within a large area environment.

while accessing additional training material. AR enables the trainee to learn the basics about the task by observing the augmented instructions and trying to perform the instructed subtask, to develop behavior and movement patterns when performing the subtasks, and to redefine those motor patterns in repeated performances of the task (i.e., during training).

Another benefit is that the trainee has real tactile feedback when performing the training task, because he or she can interact with real objects. The virtual objects provide additional information about the task and its performance and supplement the trainee's knowledge about the task. Accordingly, the trainee can access the training material (i.e., the virtual instructions, etc.) and the real environment without the need to use external separate training material (e.g., a user manual). Furthermore, the use of a training platform that involves virtual elements, as does an AR platform, allows for the measurement and evaluation of the trainee's performance in modes and levels of detail that are not possible when performing the actual task in the real world without using virtual components. By involving a technology providing virtual elements, it is also possible to respond to the trainee's performance and present corresponding feedback. In addition, the type and order of presented subtasks can be adapted, which is not possible in the real world.

A potential danger of AR applications is that users become dependent on AR features such as visual instructions. As a result, the user might not be able to perform the task when those features are not available or when the technology fails. This leads to two demands on AR training applications and programs:

- The training programs should include phases in which the amount of AR features is reduced (less virtual components, e.g., instructions only for the current subtask without additional information about the device, tools, etc.).
- The training program should also include phases in which the level of information provided by the AR features is reduced (e.g., only spatial hints without detailed instructions). That is, the level of guidance in the training system has to be adaptable to the current training phase.

To summarize, AR-based training applications must clearly differ from AR-based guiding applications, as they must really train the user and not only guide the user through the task. AR is a good technology for training, as instructions or location-dependent information can be directly linked and attached to physical objects. Because objects to be maintained usually contain a large number of similar components, the provision of location-dependent information is vitally important for the training. Furthermore, in AR-based training, sessions can be combined with tele-consultation technologies, and the availability of a trainer on-site is not mandatory.

REFERENCES

Bleser, G., and Stricker, D.: Advanced tracking through efficient image processing and visual-inertial sensor fusion. *Computers and Graphics*. 33 (2009), 1, 59–72.

Ferrin, F.J.: Survey of helmet tracking technologies. Large screen projection, avionic, and helmet-mounted displays. *Proc. SPIE*. 1456, 1 (1991), 86–94.

Hoshi, T., Okada, A., Makino, Y., Shinoda, H.: A whole body artificial skin based on cell-bridge networking system. In *Proceedings of the 3rd International Conference on Networked Sensing Systems (INSS'06)* (2006), pp. 55–60.

Kato, H., and Billinghurst, M.: Marker tracking and HMD calibration for a video-based Augmented Reality conferencing system. In *Proceedings of the Second IEEE and ACM International Workshop on Augmented Reality* (1999), IEEE Computer Society, 12.

Klein, G., and Murray, D.: Parallel tracking and mapping on a camera phone. In *Proceedings of the Eighth IEEE International Symposium on Mixed and Augmented Reality (ISMAR'09)* (2009), IEEE Computer Society, 83–86.

Kurihara, K., Hoshino, S., Yamane, K., and Nakamura, Y.: Optical motion capture system with pan-tilt camera tracking and real time data processing. In *Proceedings of the IEEE International Conference on Robotics and Automation (ICRA'02)* (2002), vol. 2, pp. 1241–1248.

Malerczyk, C.: Gestural interaction using feature classification. In *Proceedings of the Fifth International Conference on Articulated Motion and Deformable Objects (AMDO'08)* (2008), Springer, Berlin/Heidelberg, pp. 228–237.

Miller, N., Jenkins, O., Kallmann, M., and Mataric, M.: Motion capture from inertial sensing for untethered humanoid teleoperation. In *Proceedings of the Fourth IEEE/RAS International Conference on Humanoid Robots* (2004), vol. 2, pp. 547–565.

Organic Motion Inc.: Organic Motion, 2011. http://www.organicmotion.com/ (accessed October 2011).

Petersen, N., and Stricker, D.: Fast hand detection using posture invariant constraints. In *Proceedings of the 32nd Annual German Conference on Advances in Artificial Intelligence (KI '09)* (2009), Springer-Verlag, pp. 106–113.

Sudderth, E., Mandel, M., Freeman, W., and Willsky, A.: Visual hand tracking using nonparametric belief propagation. In *Proceedings of the Conference on Computer Vision and Pattern Recognition Workshop (CVPRW '04)* (2004), vol. 12, pp. 189–197.

Tanie, H., Yamane, K., and Nakamura, Y.: High marker density motion capture by retroreflective mesh suit. In *Proceedings of the IEEE International Conference on Robotics and Automation (ICRA '05)* (2005), pp. 2884–2889.

Tripicchio, P., Filippeschi, A., Ruffaldi, E., Tecchia, F., Avizzano, C.A., and Bergamasco, M.: A measuring tool for accurate haptic modeling in industrial maintenance training. In *Proceedings of the International Conference on Haptics (EuroHaptics '10)* (2010), Springer-Verlag, Berlin, pp. 377–384.

Unzueta, L., Peinado, M., Boulic, R., and Suescun, A.: Full-body performance animation with sequential inverse kinematics. *Graphical Models*, 70, 5 (2008), 87–104.

Wientapper, F., Wuest, H., and Kuijper, A.: Composing the feature map retrieval process for robust and ready-to-use monocular tracking. *Computers and Graphics*, 35, 4 (2011) 778–788.

Wuest, H., Vial, F., and Stricker, D.: Adaptive line tracking with multiple hypotheses for augmented reality. In Institute of Electrical and Electronics Engineers (IEEE): ISMAR 2005: *Proceedings of the Fourth IEEE and ACM International Symposium on Mixed and Augmented Reality (ISMAR'05)*, pp. 62–69.

Yamada, D., Maeno, T., and Yamada, Y.: Artificial finger skin having ridges and distributed tactile sensors used for grasp force control. In Proceedings of the IEEE/RSJ. *International Conference on Intelligent Robots and Systems* (2001), vol. 2, pp. 686–691.

Yoon, S.M., Malerczyk, C., and Graf, H.: Skill measurement through real-time 3D reconstruction and 3D pose estimation. In Gutiérrez, T. Sánchez, E.J. (Eds.), *LABEIN Tecnalia: SKILLS 09*: Enaction on SKILLS. DonostiaSan Sebastián: SKILLS Consortium, 2009, pp. 59–65.

7 Haptic Interfaces for Skills Training

Carlo Alberto Avizzano, Florian Gosselin,
Teresa Gutierrez, Carsten Preusche,
Emanuele Ruffaldi, Emilio Sanchez,
and Massimo Bergamasco

CONTENTS

INTRODUCTION

Multimodal virtual environments allow a fully immersive body experience for interaction and practice in digitally created contexts. They have often been proposed as relevant systems for learning (manual) procedures in a safer way, improving performances, or mastering specific skilled activities. However, the design of the multimodal interaction for training requires particular attention to ensure that at the end of the training process, the learned skills will be correctly transferred from the virtual to the real environment, and that there will not be any dependence from the training facilities. This chapter reviews how relevant haptic technologies could be designed or used for the acquisition and transfer of manual skills in training oriented systems.

TACTILE SENSING AND FINE FORCE RENDERING FOR ACQUISITION AND TRANSFER OF MANUAL TASKS

This section focuses on virtual reality (VR) platforms that help to practice surgical tasks in a repeatable and risk safe environment (Agus et al., 2002; Morris et al., 2006; Sourin et al., 2000; Terada et al., 2007). It is acknowledged that if the virtual environment is sufficiently representative of the real one, the acquired skills can be transferred to the real world (Larsen et al., 2009; Reznick and MacRae, 2006).

In particular, we address how specific haptic technologies can be designed to help in capturing and training related skills. We address simultaneously both kinesthesia and tactile sense, which are jointly employed to explore, characterize, and manipulate. The section focuses on a maxillofacial surgery (MFS) system that is described in another chapter, and on a specific surgical procedure, the Epker osteotomy. In this surgery, a corticotomy of the lower mandible is first performed: part of the body (corpus mandibulæ) is separated from the rami (rami mandibulæ). The most difficult task is to correctly drill the bone during the corticotomy. This procedure was chosen because surgeons consider it representative of delicate interventions requiring the acquisition of highly sensitive haptics-related skills. It is moreover difficult to teach and learn, as the trainees have some difficulties observing the procedure due to a narrow operating field and as trainers only very progressively let novices practice on patients; thus a very long learning process is required.

The most common errors include the following: Some young surgeons tend to be too cautious and to proceed only to a superficial and incomplete drilling, which can cause an undesirable bone fracture after drilling; other surgeons tend to grasp the instruments insufficiently firmly, producing more unstable and less controlled movements.

This justifies the development and implementation of a complete training process via haptic technologies. Here the Epker surgery is trained within a complete multimodal VR platform. Such a platform offers an alternative to real environments for the progressive acquisition of the appropriate *fine-force control* skills. It also presents relevant features such as the ability to measure the user's performance, to display quantitative feedback, to focus on critical subtasks, and to perform more repetitions in the same time period.

The training requires a realistic reproduction of the haptic information. Technical data and specification were performed through a data acquisition campaign in an anatomy laboratory. Three expert surgeons performed the complete corticotomy procedure (exposition, drilling, and distraction) on both sides of the mandibles of cadavers. Their movements (i.e., arms, hands, head), the applied forces, the accelerations of the tool, and the drilling sounds were recorded and used as a design reference for the haptic interface: workspace, peak and continuous forces/torques, maximum speed, maximum stiffness, and amplitude and frequency of vibrations.

No existing commercial haptic interface met all the task requirements. Consequently a new device development was required. A prototype of this device is shown in Figure 7.1. The control of the tool position and the force feedback are obtained thanks to a 6 DOFs robot. It makes use of a hybrid architecture composed of a 5 degrees of freedom (DOFs) parallel stage using two 3 DOFs branches connected by a link carrying an additional DOF in series. Passive parallelograms are

FIGURE 7.1 MFS training platform integrating two prototypes of the hybrid haptic device for the training of fine position and force control and visuohaptic colocation for the training of hand-eye coordination.

introduced in order to keep the orientation of the first wrist axis fixed in space and reject the singularities outside the useful workspace. As shown in Gosselin (2000), this structure combines the advantages of serial (large workspace, especially in orientation) and parallel robots (high stiffness and high transparency, especially in orientation). The ranges of motions are larger than 200 mm (close to 140°) along (around) X, Y, and Z axes. The peak force capacity is over 50 N along all axes while the control stiffness ranges from 12.4 to 20.2 N/mm at the center of the workspace (Gosselin et al., 2011).

The rendering of high-frequency drill vibrations is obtained with an active prop mounted on the robot and used as a handle. It makes use of an elongated piezoelectric stack actuator integrated in the handle and aligned with its principal axis. The bandwidth of this active prop exceeds the range of the tactile sensitivity of the human hand. The handle's surface displacements under the fingertips reach 100 μm up to 600 Hz (Giuntini et al., 2010).

The capability of this prototype to simultaneously fulfill all task requirements (i.e., 400 mm and 140° workspace, 50 N force, more than 12 N/mm stiffness, and tactile bandwidth over 600 Hz) makes it a unique tool for the study of haptically enhanced training of fine position and force control. Even if this device was developed starting from Epker performance recording, its design seems suitable for several surgical tasks, and indeed the design procedure is valid for any haptic-based VR training system.

Considering that the Epker surgery requires the capability to perform bimanual operations, a second hybrid haptic interface was integrated. Because this arm is typically used as nondominant, to execute less demanding tasks (removing surrounding tissues or cleaning the operating field), the second arm was realized with a simpler design by upgrading a preexisting master arm previously developed for abdominal telesurgery (Gosselin et al., 2005).

Finally, a fixed tangible support representing the chin was integrated as hand support in order to gain a kinesthetic reference and increase the precision of the surgeon's movements.

Humans traditionally use all their senses to gather information in the real world. Even for the Epker surgery where visual access is limited, vision still gives valuable information in addition to haptics. Analysis of this surgical task showed that it is essentially multimodal (e.g., during drilling the transition from cortical to cancellous bone is perceived by a change in bone compliance and also by a change in bone color and by a modification of the drilling sound). A VR training platform for such surgery should thus integrate visual and audio along with haptic feedback. The data acquired during the acquisition campaign show that the surgeons only slightly move their heads above the patient's mouth during the operation. Hence it is possible to implement the visual feedback with a 22 inch stereoscopic LCD monitor with active LCD shutter glasses (NVIDIA Geforce 3D Vision®) and an infrared tracking system that captures the position of the surgeon's head (NaturalPoint, TrackIR™ 5) (Gosselin et al., 2010). The 3D parameters are adjusted so that the visual feedback is coherent with the position of the tangible support and tool props. The user is far enough from the screen for comfortable vision. With this configuration, the user does not directly see his or her hands. From our experience, perfect visuohaptic colocation is not needed in this case.

The MFS training platform allows a realistic reproduction of force, tactile, audio, and visual feedback encountered during bone drilling. As shown in Chapter 13, combining subtasks focused training and sensory modulation during the training protocol (i.e., emphasis on compliance rendering via force feedback and vibrations rendering via tactile feedback before moving to the fully multimodal feedback) tends to enhance the training efficiency of the sensory motor skills required in this surgery.

On the other hand, the design of the training progression was implemented using expert data wherever relevant. For example, during the surgery task, the tool position, the applied force, and spatial 6D trajectory were used to define task profiles and characterize expertise.

The most difficult information to capture was the drill grasping force (too light for novices, resulting in unstable movements and bad quality drilling). Because sensors allowing simultaneous capture of the grasping and global forces on the tools were not available, a new sensor was developed (Nazeer, 2012) with the following features: have no task alteration; be sterilizable or removable from the sterile zone of the operating room; thin and soft (glueable to the tool or attachable to the surgery gloves); 3D measuring capabilities of both grasping and global forces; adaptable to different tools used during surgery, both small size (e.g., retractor, rugine, scalpel, etc.) and larger size (e.g., surgery drill, distraction hammer, etc.).

The proposed solution consisted in a matrix array of basic elementary sensors. Each sensor is sized between 1 and 2 mm^2 and can be integrated up to 1 cm^2 matrix. Smaller 3D sensors would be difficult to manufacture and would result in too many sensors in the 1 cm^2 matrix and thus in a too-complicated electronic. Larger sensors would not be precise enough. The sensors should be able to measure forces up to 100 N with a dynamic range of 1000 and a bandwidth above 100 Hz.

Several technologies were considered for the design of such sensors. A matrix of metallic Micro-Electro-Mechanical Systems (MEMS) strain gauges was proposed in Engel et al. (2003). It is thin, flexible, and integrates small elementary sensors. However, they do not detect shear forces. Several silicon piezoresistive sensors were also proposed. Even though they are small, such sensors are not flexible (see, e.g., Kane et al., 2000). When integrated on a flexible substrate, a larger design (Shan et al., 2005) is required. Optical sensors, based on photoelastic effect (Venketesh and Crowder, 1997) or fiber optics (De Maria et al., 2008) were also considered. Such sensors can be made flexible and 3D and fit well with the specified range of forces. However, they tend to be too large for our application. Ultrasonic acoustic sensors (Kembu et al., 2003; Shinoda et al., 1997) easily allow the identification of the shape of objects. However, they are bulky and do not allow 3D sensing. Inductive sensors (Chi and Shida, 2004; Futai et al., 2004) are too large for our applications but could be miniaturized using microelectronic technologies. They could also be implemented on flexible substrates. More recently, matrices of organic transistors on a flexible polydimethylsiloxane (PDMS) substrate were developed (Mannsfeld et al., 2010; Someya et al., 2005), as well as nanotube transistors (Takei et al., 2010). This technology is very promising but current prototypes tend to be fragile.

Our design choice was capacitive force sensors that are widely used in consumer electronics. Elementary sensors rely on a variation of the dielectric layer's thickness (Hoshi et al., 2006) or on a variation of the relative positioning of the electrodes (Chu et al., 1996). This technology allows 3D detection and can be miniaturized. Even if they require electromagnetic shielding that could limit the global sensor's flexibility, cost and specs are optimal for the chosen application.

Figure 7.2 illustrates two designs that were tested within the development. Each sensor cell is composed of three planar capacitances with values that vary according

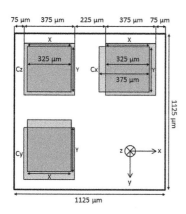
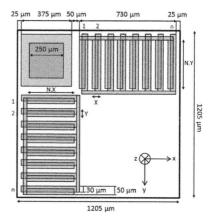

FIGURE 7.2 Example of configurations tested for the design of a miniature force sensors matrix. (Adapted from Nazeer, S., 2012.)

to the applied force. The size of one cell is slightly larger than 1 mm², allowing 8 × 8 matrices over 1 cm² for a global sensing capacity between 100 mN and 100 N. X and Y measurements are decoupled thanks to the design of the electrodes.

The design of the left is very simple. Considering a dielectric thickness of 20 μm, a deformation of 10 μm in Z corresponds to a relative variation of the capacitance of 100% produced by a force of 801 mN applied on the capacitance surface, while a deformation of 10 μm in X or Y corresponds to a relative variation of the capacitance of 3% produced by a force of 42 mN on the capacitance surface. This corresponds to 405 mN in X and Y and 7.61 N in Z on the 1 mm² sensor and to 40.5 N and 761 N on the 1 cm² matrix. This is well above our expectations. Moreover, considering a typical capacitance of 0.2 pF for this design in X and Y, 3% variation corresponds to 0.36 fF. Considering that we expect a dynamic range of 1000, this would require a resolution of 0.36 aF, which is well above the resolution of affordable electronics. A prototype was realized to validate the 3D sensing and the decoupling capabilities. This validation was made with a sensor at scale 30, which required a much lower resolution.

The design on the right in Figure 7.2 is optimized for improved sensitivity of shear forces. It is more complex. The variation of the capacitance over the whole force range is equal to 0.22 pF. The associated required capacitive resolution is 0.22 fF, which seems much more feasible. It is associated with a custom electronics based on capacitive bridges. To address matrix designs, a multiplexing electronics using discrete components was developed and tested with calibrated capacitances in the same range of values as that of the flexible sensor. The capability to measure 8 × 8 × 3 values at 25 Hz with a sensitivity of around 1 V/0.2 pF and a resolution of 0.8 mV (i.e., 0.7 fF) was validated.

FULL-BODY COORDINATION AND ATTENTION GUIDANCE THROUGH THE USE OF VIBROTACTILE INTERFACES

This section is dedicated to a type of haptic interface that, although being technologically simpler than the kinesthetic type, offers large application possibility due to the small size, variability of location on different body parts, and easy integration with other technologies. Vibrotactile interfaces are based on the fact that the human perceptual system is capable of identifying vibrations in a good range of values, allowing also a discrimination of the source of the vibration. In the context of skill training, vibrotactile devices are capable of providing spatiotemporal patterns that can be employed in different ways by varying the interpretation of the vibrotactile pattern (Baca and Kornfeind, 2006). At the most basic level, these devices can be employed for providing triggered information associated with timing of actions or for notifying the subject about error conditions. At a more advanced level, these devices can be adopted to provide feedback to the user with a continuous value, or even with directional information. A third approach focuses instead on the stimulation for learning a specific motion pattern. When the vibrotactile feedback has to affect the action immediately, it should be intuitive, as discussed by Jansen and others (2004) who presented an experiment for the selection of the positioning of vibrotactile actuators for reducing the response time in activating wrist rotations. In their case, the actuation on the external part of the hand was the most effective, corresponding to an extrinsic frame of reference with a "pull" logic.

There are several application domains that can benefit this type of interface, mainly the ones that require reduced encumbrance and increased portability. In particular, a promising domain is that of sport training, in which appropriate feedback and accelerators can be identified for improving training (Ruffaldi et al., 2011). In juggling, for example, they have been employed for supporting the rhythmic behavior of the juggling task, with a combination of external pacing and internal adaptation to subject pace. In rowing it is possible to improve both timing and motion patterns.

The rest of this section discusses some examples of application of vibrotactile devices and their perspective for future skill training research.

ACTION TIMING

This case study is related to the training of timing in rowing motion, in particular for the activation of muscles associated with the arms and back in the rowing cycle (Filippeschi et al., 2011). The aim of such training is to improve the capacity of the subjects to learn the sequence of muscle activation for optimizing the effectiveness of the rowing cycle. While previous research focused on the use of vibrotactile devices for activation feedback (van Erp et al., 2006), in this study the focus is on using an audio signal for the reference timing and then employing a vibrotactile signal to notify errors. The challenge in this application is the complexity of the motion pattern for novice subjects, due to the sequential application of different muscular sets and the distance among the stimuli.

The two reference audio signals occur at 0.67 s (back) and 1.02 s (arm) from the beginning of the stroke cycle, and, after experimental tuning, the vibrotactile feedback is used for signaling error with a symmetric threshold of 0.07 s and 0.14 s, respectively (Figure 7.3, left). For vibrotactile devices, it is necessary to decide where to apply them—in locations that are more sensitive or in the specific body part to be activated; when to apply them—right after the error or post-hoc; and finally, how to map the error to the vibrotactile signal. In this study case, one device is placed on the wrist for

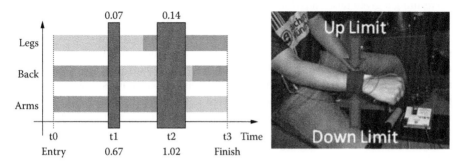

FIGURE 7.3 Left: task timing. Right: image of the vibrotactile bracelet and the chosen metaphor for motion pattern training.

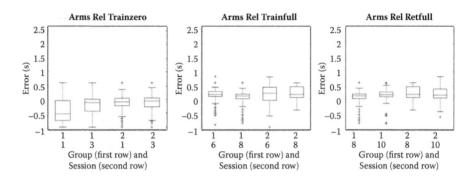

FIGURE 7.4 Statistics of training in the zero force, full force, and retention cases.

signaling the arm error, while the other is on the back. The cyclic nature of rowing motion, with a period of about 3 seconds, reduces the need for instantaneous response, but it is quite interesting to see if there is a correlation, during training, among the feedback and the performance in the next cycle. This call to action in the next cycle could be affected by the short separation between the four possible multimodal stimuli.

This approach was tested by eight subjects, organized in a between-subject design with a control group trained only with knowledge of results and audio feedback, and an experimental group trained with knowledge of results, audio, and vibrotactile feedback. The protocol considered two conditions for each subject—no force and full force—for evaluating the transfer of learned timing pattern. The dependent variables were the time of each specific activation against the reference, and the distance in time among the effective activations. The results (Figure 7.4) show that this motion pattern was quite complicated, although the training scheme is promising.

MOTION PATTERNS

A different possibility is to use vibrotactile devices for continuous augmented feedback for training motor patterns. There is a growing literature in the domain of vibrotactile vests that, by means of multiple actuators, allow stimulating the body for performing a given task. In particular, there have been efforts for defining strategies for mapping motion commands to vibrotactile commands (Spelmezan et al., 2009). The pattern adopted in this research was based on the training of the exact motion pattern of the rowing cycle, providing directional and intensity-related error information. In this design, a vibrotactile bracelet with two motors, one above the wrist and the other below, was used, taking care to reduce interferences (see Figure 7.3, right). When the motion of the hand connected to the rowing oar was too high, the upper part was vibrating proportionally to the amount of error, and vice versa. This corresponds to a push metaphor following previous research (Ruffaldi et al., 2009).

PERSPECTIVES

The adoption of vibrotactile feedback for skill training is a proven technology for basic and single vibrotactile stimulation, while it requires further investigation for understanding the effectiveness of multiple information flowing by different vibrotactile actuators. There are several interesting open questions in the domain of multimodal integration, in particular for estimating, in the long term, the ability of vibrotactile feedback for improving retention of the learned motor pattern.

HAND, EYE, AND HAPTICS COORDINATION IN THE USE OF ENCOUNTERED HAPTICS

Most haptic devices act through a stable contact with the user, like in the Phantom device (Massie, 1994), or larger exoskeletons (Bergamasco, 2007). Instead, encountered-type haptic (EH) interfaces allow the user to not always be in contact with the interface. The device encounters the user only when he or she is going to be in contact with a virtual object (McNeely, 1993; Tachi, 1994).

EH interfaces provide advanced capabilities for training in the cases where the coordination among the user body and external objects is concerned (pointing, ball playing, object catching, etc.). The *noncontinuous contact* nature of these systems allows the natural motion dynamic of the user body to be minimally affected. In terms of skills learning, this provides a huge advantage because it allows for the re-creation of a simulated environment that is closer to the real one, and therefore minimally distorts the training conditions during virtual interaction.

Current research mostly focuses on mechanical design issues (Gruenbaum, 1997; McNeely, 1993; Yokokohji, 1996), the proper surface contact rendering in the inter-action with plates (Solazzi, 2006), or manipulation through patches (Shigeta, 2007; Yokokohji, 2005) or strips (Bordegoni, 2008). Here we address a complementary view that is in the use of these devices for specific coordination issues that include motion learning and control. In particular, a notion of dual control-simulation among graphi-cal entities and the haptic interfaces will be addressed as well as the requirements for spatial and kinematic coherence only during contact interaction. When properly addressed, these properties extend the limit of interaction between user and virtual environment beyond the limitation given by their mechanical workspace. In addition we will consider, in contrast with slow motion contact referred to in current research, the issues relating the dynamics of hand and object interaction. In skills training, this dynamic is critical for the proper representation of hand/eye coordination and learning.

Adding dynamical properties to EH introduces a set of new control and percep-tion issues to keep under consideration for a proper control design and perception modeling. A comparison with traditional EH design is showed in the following table:

Static Issues	Dynamic Issues
Reachability, contact orientation, local geometry, contact prediction, rendering of contact stiffness, multipoint grasping, plates interference	Mass/weight/inertia properties and compensation effects, cinematic of motion and related distortion, reachable workspace, impact prediction, motion constraints

Accurate representation of the *mass/weight/inertia* properties during interaction with the user is fundamental in order to allow that the force reflected during the training is showing conditions that can be correlated during the execution of real tasks. The control of these properties is possible by means of feed-forward strategies (if the trajectory is known ahead) and closed-loop feedback starting from the Lagrange formulation (Sciavicco, 2003):

$$M(q)\ddot{q} + C(q,\dot{q})\dot{q} + \Delta_{visc}\dot{q} + G(q) = Q + J^T F_c$$

where M, C, G, Δ are nonlinear functions describing the interface inertia properties; Q is the torque contribution delivered by motors on the interface jonts; J is the Jacobian at the contact point; and F_c is the user contact force. By correlating this equation to the mass control of the theoretical manipulation model,

$$F_c = M_{obj}(g + \ddot{x})$$

We achieve an ideal control strategy expressed as

$$Q = M(q)\ddot{q} + C(q,\dot{q})\dot{q} + D_{visc}\dot{q} + G(q) - J^T M_{obj}(g + J(q)\ddot{q} + \frac{dJ(q)}{dt}\dot{q}$$

which is usually simplified according to specific system/application configurations. When the device is ungrasped, the robot control is free to adopt the optimization strategy that is more convenient for task control (such as the workspace optimization, energy saving, user safety, etc.). However, it is fundamental to predict the next contact instant and to regulate free motion trajectories to match specific impact conditions at that moment. We determined these conditions through experimental analyses.

The description of this experimental analysis is given in Tripicchio (2009), where the 3-DOF device shown in Figure 7.5 and described in Avizzano (2005) was used to control a tennis ball falling into the user's hand. The experimental setup was organized to compare the perception of real balls falling into the hand with the perception produced by the robot-controlled balls. Several equivalence conditions were investigated taking into account kinematic, static, and dynamic properties of the haptic interface, the real ball and the virtual object. Two basic equivalence models were finally selected: equivalence determined on the basis of the energy loss (between virtual and real environment), and equivalence determined on the basis of the pure impact velocity (isokinetic).

The experiments performed have shown that the first class condition is more correlated to perception than the second, devising in such a way a particular contact condition in which the haptic device has to impact the hand at a different velocity than the virtual object: $v_{hi} = k\,v_{obj}$, where the scaling velocity factor k is determined as a function of pre- and post-contact conditions (Tripicchio, 2009) using the reflected Cartesian inertia of the device at the contact point (Khatib, 1986). Another important aspect of EH is the ability to predict the proper position and time of the next collision. This operation is analogous to the prospective control (Turvey, 1992) that humans use for affordances and to predict appropriate behavior in the environment interaction. The existence of a precise physical

FIGURE 7.5 The encountered ball interface. (From PERCRO, 2010. With permission.)

model of the environment allows us to reduce this problem to appropriate monitoring of user hands and to predict collision time with first- or second-order hand motion interpolation.

Motion constraints and trajectory training can be taught using virtual fixture techniques that are triggered once the contact is established. Typically a damped virtual stiffness orthogonal to the constraint has proven to be of enough quality to ensure the adherence to the trajectory. Techniques from robotic rehabilitation (Satler, 2011) allow also the decoupling of the physical behavior between the constrained axes and the physically controlled ones. In all the above cases, the additional *apparent forces* terms caused by the virtual constraint should be added in feed-forward to compensate for the nonlinear geometry of the constraint.

Time versus velocity profiles have been assessed during juggling virtual balls (Ruffaldi et al., 2011). People were able to juggle with the virtual balls using different factors of gravity acceleration with minimal distortion of performed trajectories.

BIMANUAL HAPTIC SYSTEMS FOR MOTOR SKILLS LEARNING

Bimanual coordination is relevant for several manipulation tasks where one hand is not enough. The Bimanual Haptic Desktop System (BHDS) is a large dual-contact haptic device specifically designed to simulate bimanual force coordination tasks. The primary application of the system is for neurorehabilitation, but it could also be applied to other tasks. The BHDS, as shown in Figure 7.6, possesses four independent degrees of freedom and allows arms to move on a plane while subjected to different force constraints.

The simultaneous control of two arms allows several correlated motion control strategies that solicit left-to-right or right-to-left cross mapping to be introduced. Basically the two arms can move in the same workspace while controlled with compliance/stiffness/force and viscosity loops. In addition, the coordination control may

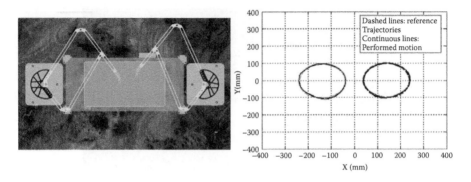

FIGURE 7.6 BHDS design and control structure and constrained circular motion task.

impose a common action policy that correlates the type of stimuli and the response felt by both arms (Figure 7.7).

Different types of controllers can be implemented to guide the two arms, depending on how much transparency is required among them. In a first option, no transparency desired, one of the two arms (named the "master") is left free to move, while the other (the "slave") is guided with a pure position control that imposes the second hand to follow motions that are guided only by the first hand. A second option is to leave the second hand free to perform small deviations from the reference trajectory generated by the master, hence implementing an impedance controller on the second arm which determines the level of attraction the first arm has to impose to the second. As a third option, it is possible to impose different levels of force reflection among the two arms by using bilateral impedance models that allow each arm to show part of the forces felt by the second arm. Finally, it is possible to constrain the guiding modalities described above along given trajectories in the operational workspace, and in such a way the guiding control maps from a 2D problem into a 1D one. In Figure 7.7 an example of constrained circular motion is shown. These guiding modalities allow the device to move along any type of 1D geometric curve using only linear coordinates. When the guidance is not constrained, apart from the case in which the two arms are considered to be independent, there are several possible mapping strategies to coordinate the motion of the two arms. In these cases we assume that, apart from a reference offset that can be arbitrarily established among the two arms, a specific differential motion constraint is held within the motion.

There are several guidance opportunities that solicit specific differential motion properties:

- *Isokinetic motion*: The second arm replicates exactly the motion of the first with a pure offset difference (application to manipulation). In this type of motion one hand governs the motions of the other which is tracking with a proper offset the same geometries described by the first hand. The coordination property is useful to transfer gestures across right and left spheres of the users, for instance in the cases when one skill such as writing is correctly learned by one side and needs to be mapped to the other one.

LR map	Property	Control model
Isocinematic	$X_1(t_2) - X_1(t_1) = X_2(t_2) - X_2(t_1)$ $Y_1(t_2) - Y_1(t_1) = Y_2(t_2) - Y_2(t_1)$	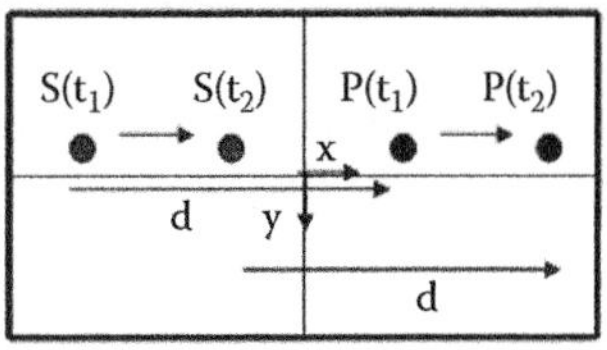
X or Y symmetry	$X_1(t_2) - X_1(t_1) = X_2(t_1) - X_2(t_2)$ $Y_1(t_2) - Y_1(t_1) = Y_2(t_2) - Y_2(t_1)$ $X_1(t_2) - X_1(t_1) = X_2(t_2) - X_2(t_1)$ $Y_1(t_2) - Y_1(t_1) = Y_2(t_1) - Y_2(t_2)$	
O symmetry	$X_1(t_2) - X_1(t_1) = X_2(t_1) - X_2(t_2)$ $Y_1(t_2) - Y_1(t_1) = Y_2(t_1) - Y_2(t_2)$	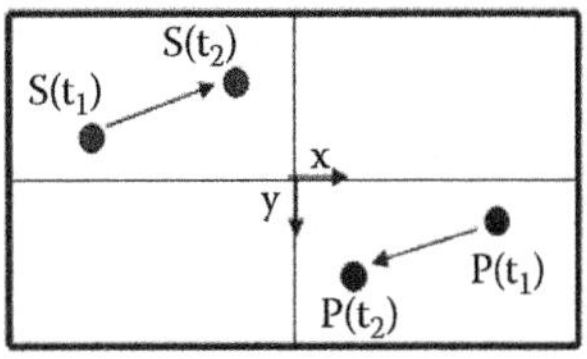
Scaled motion	$X_1(t_2) - X_1(t_1) = k_x(X_2(t_2) - X_2(t_1))$ $Y_1(t_2) - Y_1(t_1) = k_y(Y_2(t_2) - Y_2(t_1))$	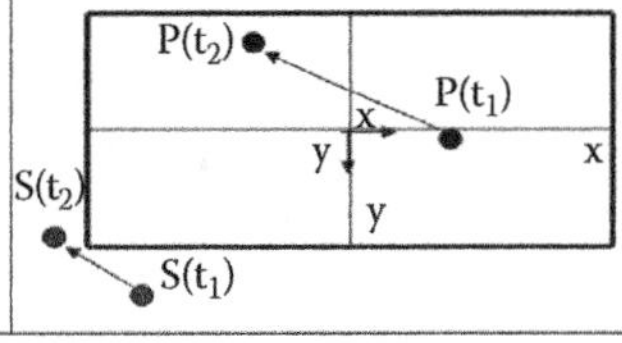

FIGURE 7.7 Motion guidance styles.

- *Axial symmetric motion*: One axis is inverted such that the differential motion of the two arms is symmetric with respect this axis (grasping). This mapping policy acts like a mirror—all trajectories performed by the first arm are repeated by the other. The strategy can be used in different contexts where the two hands cooperate to execute a specific task, such as lifting an object.
- *Point symmetric motion*: Both axes are inverted to achieve the reference position of the second arm. Point symmetry is useful when objects or geometries are replicated along a circle (a sample motion is described by car driving). Point symmetric motions are useful to keep into account operations that consider the full extension of one arm together with the rotation of the user's torso. This particular kind of mapping allows the guidance motion closer to the user's body while the guided one is not as close.
- *Free scaled motion*: One arm performs the motion of the second one in a smaller (or larger) area, thus allowing the second arm to perform the exercises with a larger workspace (or in a more accurate way).

Other (nonpurely kinematic) policies include the possibility to insert soft or variable constraints between the two arms beside the bilateral (tele-operation) model described before; it is not uncommon to use motion penalty models in which the task achievement is performed with smaller kinematic displacement when performed from one of the two arms, and force penalty models that require asymmetric force levels for performing the same operation.

VIRTUAL FIXTURES GUIDANCE FOR THE TRANSFER OF COGNITIVE AND MOTOR SKILLS

The first aim of haptic technology is to realistically render forces in the virtual world, but it can also be used to directly help trainees understand and perform a target task by the use of virtual fixtures. The term *virtual fixture* (VF) is most often used to refer to a task-dependent aid that guides the user's motion along desired directions while preventing motion in undesired directions or regions of the workspace. Generally speaking, a VF can be any information transmitted to the human by means of any sensory channel. Thus, a VF can be rendered with graphics (visual VF), with force features (haptic VF), and with audio signals (audio VF). This section focuses on haptic VFs.

Some research works are showing promising potential in the use of VFs for fostering skill acquisition (Kuang et al., 2004; Prada and Payandeh, 2005; Vallery et al., 2009; Zheng et al., 2006). However, special attention must be paid to the design of training protocols that use haptic VFs to avoid disturbing the skill acquisition and to guarantee the skills transfer to the real world, where these VFs are not available. In this sense, it is important that the effects of the VFs can be gradually changed or reduced, displaying at each time adequate information according to the trainees' level of expertise and their evolution along the training process.

The SKILLS consortium has developed a haptic VFs library that can be used for the training of different cognitive and motor skills such as fine motor control (to perform fine, precise, and delicate movements) or bimanual coordination. It is focused on impedance-type VFs, although it could be adapted to admittance-type haptic rendering. The output of these VFs is force and torque dependent on the trainees' motion and the type of VF (e.g., trainees can receive forces to constrain their motion along a specific direction or just attraction/repulsion forces with different levels of intensity).

A VF in this library is defined by a geometric primitive and a force profile. The following primitives can be used to design different types of motion guidance and skill transfer support:

- Point in space: Corresponds to a virtual guide in the form of an attractor or a repulsing point (see Figure 7.9). It can be used, for example, to guide the trainee toward the correct position to perform an insertion in a surgical task, or to prevent the trainee from entering the spherical volume around a damageable device.

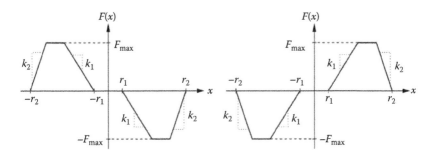

FIGURE 7.8 Force profile of a VF described by the inner and outer radii (r_1/r_2), the stiffness (k_1/k_2), and the maximum force (F_{max}). Left: attracting VF; right: repulsing VF.

- Trajectory in space: Pulls or pushes the trainee from the closest point on a trajectory that would represent an ideal motion. It can be used, for example, to teach an assembly path that avoids damages in the machines.
- Trajectory in time: Allows the trainee not only to follow a specific trajectory but also to do it with a specific speed.
- Plane or surface: Pulls or pushes the trainee from the closest point on a plane or surface. It can be a guide toward the working region (where actions happen) or to indicate the border of a forbidden region. For example, a virtual wall prevents the trainee from entering into a part of the workspace, and a haptic canyon pushes the arm of a robot toward the preferred area defined.

In the abstraction of the haptic VFs, a general force profile has been defined by means of the following parameters (see Figure 7.8):

- Two radii describing the inner and outer bounds of the area where the VF is defined (i.e., the fixture will not have any effect out of this area)
- The stiffness defining the scope of force characteristics (dependent from the haptic device)
- The upper limitation for the force (dependent from the haptic device)
- The type of force: attraction or repulsion
- The intensity of the VF: from the maximum value "1" to the minimum one "0" (in this case there will not be any force, it is like if the fixture does not exist)

Summary

A haptic VF can be given in the joint or the task space, can be attractive or repulsive, and can be restricted to a certain area or be active in the whole area, as can be seen in Figure 7.9 that shows a point VF. Combining the different alternatives for the geometric primitives and the force profile parameters a wide variety of haptic VFs can be generated. Each trainer should select the best option for each training scenario according to the type of the target task and the skills to be learned and the trainee's profile.

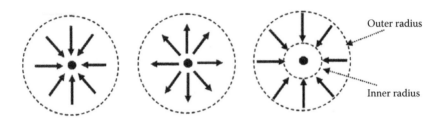

FIGURE 7.9   Virtual fixture corresponding to a point. Left: attracting. Middle: repulsing. Right: active area of the virtual fixture.

REFERENCES

Agus, M., Giachetti, A., Gobbetti, E., Zanetti, G., and Zorcolo, A.: Real-time haptic and visual simulation of bone dissection. In: *Proceedings of IEEE Virtual Reality*, Orlando, Florida, pp. 209–216 (2002).

Avizzano, C.A., Marcheschi, S., and Bergamasco, M.: 13 Haptic interfaces: Collocation and coherence issues. *Multi-point Interaction with Real and Virtual Objects*, 201–213. Springer (2005).

Baca, A., and Kornfeind, P.: Rapid feedback systems for elite sports training, *Pervasive Computing* 5(4), 70–76 (2006).

Bergamasco, M., Frisoli, A., and Avizzano, C.: Exoskeletons as man-machine interface systems for teleoperation and interaction in virtual environments. *Advances in Telerobotics*, 61–76. Springer (2007).

Bordegoni, M., Ferrise, F., Shelley, S., Alonso, M.A., and Hermes, D.: Sound and tangible interface for shape evaluation and modification. *Haptic Audio visual Environments and Games, 2008. HAVE 2008. IEEE International Workshop on* (pp. 148–153) (2008).

Chi, A., and Shida, K.: A new multifunctional tactile sensor for three-dimensional force measurement, *Sensors and Actuators A: Physical* 111(2–3), 172–179 (2004).

Chu, A., Sarro, P.M., and Middelhoek, S.: Silicon three-axial tactile sensor, *Sensors and Actuators A, Physical* 54(1–3), 505–510 (1996).

De Maria, G., Minardo, A., Natale, C., Pirozzi, S., and Zeni, L.: Optoelectronic tactile sensor based on micromachined scattering wells, *First Mediterranean Photonics Conference*, Ischia, Italy (2008).

Engel, J., Chen, J., Liu, C., Flachsbart, B.R., Selby, J.C., and Shannon, M.A.: Development of polymide-based flexible tactile sensing skin (2003).

Filippeschi, A., Ruffaldi, E., and Korman, M.: Preliminary evaluation of timing training accelerator for the sprint rowing system. In B.G. Bardy, J. Lagarde, and D. Mottet (Eds.), *Proceedings of the 2011 International SKILLS Conference*, Montpellier, France, December 15–16 (2011).

Futai, N., Matsumoto, K., and Shimoyama, I.: A flexible micromachined planar spiral inductor for use as an artificial tactile mechanoreceptor, *Sensors and Actuators. A, Physical*, 111(2–3), 293–303 (2004).

Giuntini, T., Ferlay, F., Bouchigny, S., Gosselin, F., Frisoli, A., and Beghini, M.: Design of a new vibrating handle for a bone surgery multimodal training platform. In *Proceedings of the 12th International Conference on New Actuators*, Bremen, Germany, pp. 85–88 (2010).

Gosselin, F.: Développement d'outils d'aide à la conception d'organes de commande pour la téléopération à retour d'effort, PhD dissertation, Poitiers University, France (2000).

Gosselin, F., Bidard, C., and Brisset, J.: Design of a high fidelity haptic device for telesurgery, *Proceedings of the IEEE International Conference on Robotics and Automation*, Barcelona, Spain, April 18–22, pp. 206–211 (2005).

Gosselin, F., Ferlay, F., Bouchigny, S., Mégard, C., and Taha, F.: Design of a multimodal VR platform for the training of surgery skills, *Haptics: Generating and Perceiving Tangible Sensations, LNCS 6192, Part II, Proceedings of the International Conference on EuroHaptics,* Amsterdam, the Netherlands, 8–10 July, pp. 109–116 (2010).

Gosselin, F., Ferlay, F., Bouchigny, S., Mégard, C., and Taha, F.: Specification and design of a new haptic interface for maxillo facial surgery, *Proceedings of the IEEE International Conference on Robotics and Automation*, 9–13 May, Shanghai, China, pp. 737–744 (2011).

Gruenbaum, P.E., and McNeely, W.: Implementation of dynamic robotic graphics for a virtual control panel. *Presence: Teleoperators and Virtual Environments*, 6(1), 118–126 (1997).

Hoshi, T., Okada, A., Makino, Y., and Shinoda, H.: A whole body artificial skin based on cell-bridge networking system, *Proc. Third International Conference on Networked Sensing Systems (INSS 2006)*, May, Rosemont, Illinois, pp. 55–60 (2006).

Jansen, C., Oving, A., and Van Veen, H.: Vibrotactile movement initiation. *Proceedings of Eurohaptics*, pp. 110–117 (2004).

Kane, B.J., Cutkosky, M.R., and Kovacs, G.T.A.: A piezoresistive three-axial tactile sensor element made by surface micromachining, *Sensor* (2000).

Kembu, T., Masachika, I., and Keigo, W.: Principal curvature measurement by an acoustic tactile sensor, *Transactions of the Japan Society of Mechanical Engineers, C*, 69(684), 2057–2063 (2003).

Khatib, O., and Burdick, J.: Motion and force control of robot manipulators. *Robotics and Automation. Proceedings. 1986 IEEE International Conference on* (Vol. 3, pp. 1381–1386) (1986).

Kuang, A.B., Payandeh, S., Zheng, B., Henigman, F., and MacKenzie, C.L.: Assembling virtual fixtures for guidance in training environments, *Haptic Interfaces for Virtual Environment and Teleoperator Systems, International Symposium on; 12th International Symposium on Haptic Interfaces for Virtual Environment and Teleoperator Systems (HAPTICS'04)*, pp. 367–374 (2004).

Larsen, C.R., Soerensen, J.L., Grantcharov, T.P., Dalsgaard, T., Schouenborg, L., Ottosen, C., Schroeder, T.V., and Ottesen, B.S.: Effect of virtual reality training on laparoscopic surgery: Randomized controlled trial, *BMJ* 338, b1802 (2009).

Mannsfeld, S.C.B., Tee, B.C.K., Stoltenberg, R., Chen, C.V.H.H., Barmann, S., Muir, B.V.O., Sokolov, A.N., Reese, C., and Bao, Z.: Highly sensitive flexible pressure sensors with micro-structured rubber as the dielectric layer, *Nature Materials* 9, 859–864 (2010).

Massie, T.H., and Salisbury, J. K.: The phantom haptic interface: A device for probing virtual objects. *Proceedings of the ASME winter annual meeting, symposium on haptic interfaces for virtual environment and teleoperator systems* (Vol. 55, pp. 295–300) (1994).

McNeely, W.A.: Robotic graphics: A new approach to force feedback for virtual reality. *Virtual Reality Annual International Symposium, 1993, 1993 IEEE* (pp. 336–341) (1993).

Morris, D., Sewell, C., Barbagli, F., Salisbury, K., Belvins, N.H., and Girod, S.: Visuohaptic simulation of bone surgery for training and evaluation, *Computer Graphics and Applications* 26(6), 48–57 (2006).

Nazeer, S.: Conception et réalisation de micro-capteurs de pression pour l'instrumentation d'interface à retour d'effort, Ph.D. dissertation (in French), University of Paris-Sud 11, Orsay, France (2012).

Palluel-Germain, R., Bara, F., de Boisferon, A.H., Hennion, B., Gouagout, P., and Gentaz, E.: A visuo-haptic device—Telemaque—increases kindergarten children's handwriting acquisition. In: *Proceedings of the Worldhaptics Conference*, Tsukuba, Japan, pp. 72–77 (2007).

Prada, R., and Payandeh, S.: A study on design and analysis of virtual fixtures for cutting in training environments, *Eurohaptics Conference, 2005 and Symposium on Haptic Interfaces for Virtual Environment and Teleoperator Systems, 2005. World Haptics 2005. First Joint*, March 18–20, pp. 375–380 (2005).

Reznick, R.K., and MacRae, H.: Teaching surgical skills, changes in the wind, *New England Journal of Medicine* 355(25), 2664–2669 (2006).

Ruffaldi, E., Filippeschi, A., Frisoli, A., Sandoval-Gonzalez, O., Avizzano, C. Q., and Bergamasco, M.: Vibrotactile perception assessment for a rowing training system, *Proceedings of the Third IEEE Joint Conference on Haptics, World Haptics* (2009). doi: http://doi.ieeecomputersociety.org/10.1109/WHC.2009.4810849.

Ruffaldi, E., Filippeschi, A., Avizzano, C.A., Bardy, B., Gopher, D., and Bergamasco, M.: Feedback, affordances and accelerators for training sports in virtual environments, *MIT Presence* 20(1) (2011).

Ruffaldi, E., Tripicchio, P., Avizzano, C.A., and Bergamasco, M.: Haptic rendering of juggling with encountered type interfaces. *PRESENCE: Teleoperators and Virtual Environments*, 20(5), 480–501. MIT Press (2011).

Satler, M., Avizzano, C.A., and Ruffaldi, E.: Control of a desktop mobile haptic interface. *World Haptics Conference (WHC), 2011 IEEE* (pp. 415–420) (2011).

Sciavicco, L., and Siciliano, B.: *Modelling and control of robot manipulators*. Springer Verlag, London (2000).

Shan, J.H., Mei, T., Sun, L., Kong, D.Y., Zhang, Z.Y., Ni, L., Meng, M., and Chu, J.R.: The design and fabrication of a flexible three-dimensional force sensor skin, *Proceedings of the IEEE/RSJ International Conference on Intelligent Robots Systems (IROS 2005)*, pp. 1965–1970 (2005).

Shigeta, K., Sato, Y., and Yokokohji, Y.: Motion planning of encountered-type haptic device for multiple fingertips based on minimum distance point information. *EuroHaptics Conference, 2007 and Symposium on Haptic Interfaces for Virtual Environment and Teleoperator Systems. World Haptics 2007. Second Joint* (pp. 188–193) (2007).

Shinoda, H., Matsumoto, K., and Ando, S.: Tactile sensing based on acoustic resonance tensor cell, *International Conference on Solid-State Sensors and Actuators,* Chicago, June (1997).

Solazzi, M., Frisoli, A., Salsedo, F., and Bergamasco, M.: An innovative portable fingertip haptic device. *ENACTIVE/06*, 229. Citeseer (2006).

Someya, T., Kato, Y., Sekitani, T., Iba, S., Noguchi, Y., Murase, Y., Kawaguchi, H., and Sakurai, T.: Conformable, flexible, large-area networks of pressure and thermal sensors with organic transistor active matrixes, *PNAS* 102, 12321–12325 (2005).

Sourin, A., Sourina, O., and Howe, T.S.: Virtual orthopedic surgery training, *Computer Graphics and Applications* 20(3), 6–9 (2000).

Spelmezan, D., Jacobs, M., Hilgers, A., and Borchers, J.: Tactile motion instructions for physical activities, *Proceedings of the 27th International Conference on Human Factors in Computing Systems*, pp. 2243–2252 (2009).

Tachi, S., Maeda, T., Hirata, R., and Hoshino, H.: A construction method of virtual haptic space. *Proceedings of the 4th International Conference on Artificial Reality and Tele-Existence (ICAT'94)* (pp. 131–138) (1994).

Takei, K., Takahashi, T., Ho, J.C., Ko, H., Gillies, A.G., Leu, P.W., Fearing, R.S., and Javey, A.: Nanowire active-matrix circuitry for low-voltage macroscale artificial skin, *Nature Materials* 9(10), 821–826 (2010).

Terada, T., Ogata, M., Kukikawa, T., Hongo, S., Nagasaka, M., Takanami, K., Kajihara, K., and Fujino, M.: Virtual human body using haptic devices for endoscopic surgery training simulator. In *Proceedings of the Fourth IEEE International Conference on Mechatronics*, Kumamoto, Japan, pp. 1–5 (2007).

Tripicchio, P., Ruffaldi, E., Avizzano, C.A., and Bergamasco, M.: Control strategies and perception effects in co-located and large workspace dynamical encountered haptics. *EuroHaptics conference, 2009 and Symposium on Haptic Interfaces for Virtual Environment and Teleoperator Systems. World Haptics 2009. Third Joint* (pp. 63–68) (2009).

Turvey, M.T.: Affordances and prospective control: An outline of the ontology. *Ecological Psychology*, 4(3), 173–187. Taylor & Francis, Boca Raton, FL (1992).

Vallery, H., Guidali, M., Duschau-Wicke, A., and Riener, R.: Patient-cooperative control: Providing safe support without restricting movement. In *Proceedings of World Congress on Medical Physics and Biomedical Engineering*, Munich, Germany, September 7–12, pp. 166–169, Springer Verlag (2009).

van Erp, J.B.F., Saturday, I., and Jansen, C.: Application of tactile displays in sports: Where to, how and when to move, *Proceedings of the Eurohaptics International Conference*, pp. 3–6 (2006).

Venketesh, N.D., and Crowder, R.M.: A dynamic tactile sensor on photoelastic effect, *Sensors and Actuators A* 128(2006), 217–224 (1997).

Yokokohji, Y., Hollis, R.L., and Kanade, T.: What you can see is what you can feel-development of a visual/haptic interface to virtual environment. *Virtual Reality Annual International Symposium, 1996., Proceedings of the IEEE 1996* (pp. 46–53) (1996).

Yokokohji, Y., Muramori, N., Sato, Y., and Yoshikawa, T.: Designing an encountered-type haptic display for multiple fingertip contacts based on the observation of human grasping behaviors. *The International Journal of Robotics Research*, 24(9), 717–729. Sage (2005).

Zheng, B., Kuang, A., Henigman, F., Payandeh, S., Lomax, A., Swanström, L., and MacKenzie, C.L.: Effects of assembling virtual fixtures on learning a navigation task. In *Medicine Meets Virtual Reality 14: Accelerating Change in Healthcare: Next Medical Toolkit*, IOS Press, pp. 586–591 (2005).

Zheng, B., Kuang, A., Henigman, F., Payandeh, S., Lomax, A., Swanström, L., and MacKenzie, C.L.: Effects of assembling virtual fixtures on learning a navigation task. *Studies in Health Technology and Informatics*, pp. 586–591 (2006).

COLOR FIGURE 5.6 Cluster of markers employed in the proposed video-based motion capture system.

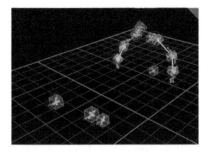

COLOR FIGURE 5.8 Motion capture scenario with clusters of markers attached to both the rowing machine and the athlete.

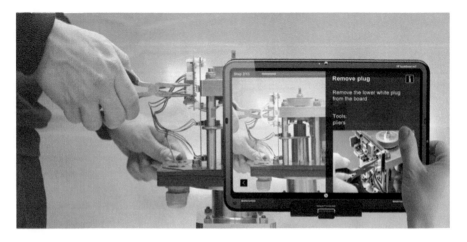

COLOR FIGURE 6.1 Augmented reality training of an assembly task.

COLOR FIGURE 6.5 Distributed AR training session with mobile trainee and remote trainer.

COLOR FIGURE 6.6 Augmented reality training within a large area environment.

COLOR FIGURE 11.1 The SPRINT system.

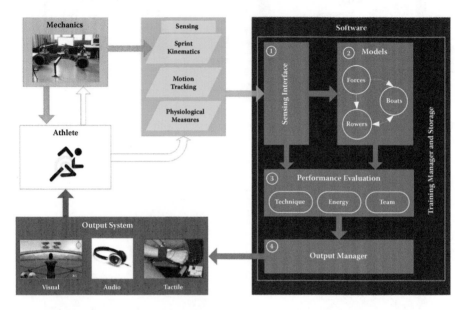

COLOR FIGURE 11.2 Architecture of the SPRINT rowing system.

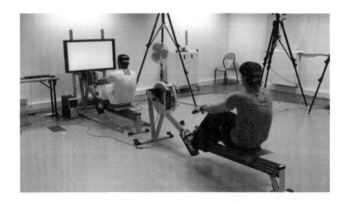

COLOR FIGURE 11.7 Experimental setup. Coordination with real teammate.

COLOR FIGURE 12.1 A view of the Light Weight Juggler (LWJ).

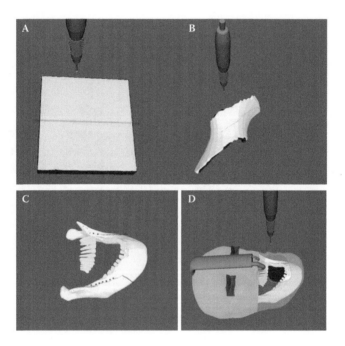

COLOR FIGURE 13.1 Example of exercises: (A) drilling a line on a flat surface (SK protocol); (B) drilling a line on a curved surface (SK); (C) line of the Epker osteotomy seen on a maxilla (SI); (D) drilling the cortical with tissue interaction (last exercise for both SK and SI).

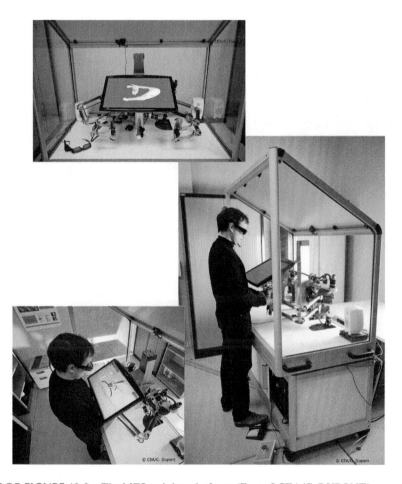

COLOR FIGURE 13.2 The MFS training platform. (From ©CEA/C. DUPONT.)

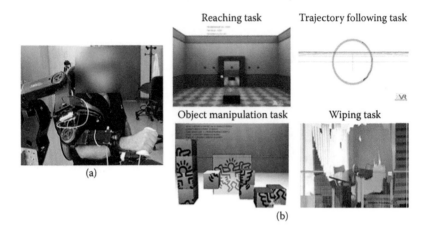

Reaching task Trajectory following task

Object manipulation task Wiping task

(a)

(b)

COLOR FIGURE 14.1 (a) The Light-Exoskeleton Platform; (b) some training exercises proposed with the system.

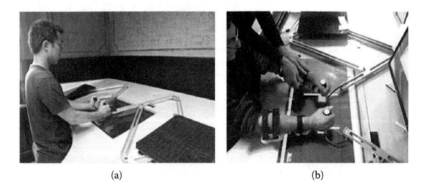

(a) (b)

COLOR FIGURE 14.11 Bimanual Haptic Desktop System (a) and (b) an example of one subject doing a simulated lifting task.

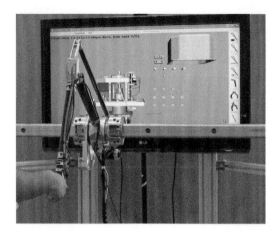

COLOR FIGURE 15.2 IMA-VR training platform: The trainee interacts and manipulates the virtual scene combining haptic, audio, and visual feedback.

COLOR FIGURE 15.4 IMA-AR platform. Left: The trainee interacts with the IMA-AR platform using a tablet PC that acts as a see-through device; additional haptic feedback is presented via a vibrotactile bracelet. Right: The interactive training application running on the tablet PC.

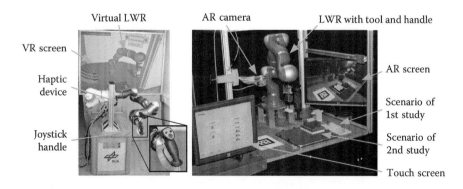

COLOR FIGURE 16.1 The two training setups. Singularity training setup (left) and compliance training setup (right).

8 Auditory Interfaces in Virtual Environment Training

Pablo F. Hoffmann and Dorte Hammershøi

CONTENTS

INTRODUCTION

The various advantages of using virtual environments for training have been emphasized in previous chapters. In addition, studies primarily concerned with learning have consistently shown the benefits of multisensory learning (Shams and Seitz, 2008). Because human interaction with the real world often, if not always, involves multisensory stimulation, it seems reasonable to argue that multimodal interfaces in virtual environment training must be the rule. However, in the process of designing multimodal interfaces for training, we must bear the importance of evaluating the significance of the individual modalities so that we can maximize the effectiveness of the multimodal virtual environment for training. In plain terms, we should use each modality for what it does best.

In this chapter we discuss the potential role of auditory interfaces in virtual training systems. Aspects involving facilitation of skill acquisition by complementing and supplementing other sensory modalities are examined. In discussing how auditory interfaces may best be employed to support training in virtual environments, we define auditory interfaces in the same line as the more general concept of *auditory display* (i.e., as the application of sounds in human-machine interfaces with the purpose of conveying *meaningful* information). We highlight the keyword *meaningful* in the sense that the information displayed via the auditory channel must prove valid for skill acquisition and performance enhancement. Its validity should be reflected by significant changes in quantitative variables such as time to complete a task or magnitude or probability of error, as well as qualitative variables such as perceived workload or situational awareness (Hale et al., 2009).

Auditory interfaces have the potential to increase the applicability of virtual environments in training by reinforcing attributes such as immersion, usability, and multimodality (Novo, 2005). These are all important factors bearing direct implications to the quality and extent to which the skills acquired during virtual training can be transferred to the real world. At a general level, research on sound perception suggests that sound can be an effective channel of information for improving performance in temporal and spatial processing skills. For temporal processing skills, auditory information has been shown to be effective at improving performance on tasks in which rhythm or timing is critical to the performance of a sensory-motor skill (e.g., juggling, rowing, or swimming) (Wang and Hart, 2005). For spatial processing skills, we can mention tasks including navigation (Gonot, Chateau, and Emerit, 2006; Lokki and Gröhn, 2005), and mobility and orientation for the blind (Merabet and Sánchez, 2009; Seki and Sato, 2011). Recent efforts have been devoted to evaluate the potential of auditory interfaces based on various sound dimensions such as pitch, loudness, harmonicity, and sound direction for the training of complex sensory-motor skills (Sigrist et al., 2011), gait rehabilitation (Wellner et al., 2008), assembly tasks (Zhang, Xiao, and Travis, 2006), tactile enhancement targeting surgical tasks (Müller-Tomfelde, 2004), and conveying abstract knowledge in environments we cannot access in real life (Garcia-Ruiz and Gutierrez-Pulido, 2006). A theoretical model based on temporal and spatial processing information has recently been proposed in the form of guidelines for the design, integration, and evaluation of auditory interfaces for interactive applications (Ahmad, Stanney, and Fouad, 2009). These guidelines target user-centered auditory interfaces and attempt to capitalize on the advantages of using the auditory modality for improving performance and facilitating acquisition of skills associated with temporal and spatial information processing, and how both forms of processing can be integrated into a unified framework.

The topics we consider in this chapter involve skill acquisition in application contexts that emphasize sensory-motor responses as well as the significance of multimodal interactions that may affect skill acquisition. In addition to presenting the actual applications, the technological aspects of implementing auditory interfaces, and their integration in multimodal systems, are generally described.

AUDITORY INTERFACES IN MULTIMODAL VIRTUAL ENVIRONMENTS

In virtual environment systems, multimodal architectures may be based on an event-driven paradigm to generate parameters that control rendering of the different modalities (e.g., collision detection) (Avanzini and Crosato, 2006b). For example, in a collision event, parameters such as size, texture, and shape of the colliding objects may be calculated for the visual modality. During collision, these visual parameters may in turn be used to derive acoustics parameters, such as attack time, decay time, and timbre. Likewise, this can be extended to the spatial position of the objects in the virtual environment. In dynamic conditions, for example, the spatial correspondence between the user and the virtual objects must be constantly updated by the graphical engine to change the visual display accordingly. The same information can be used to drive the sound engine, by providing spatial coordinates to control sound direction.

Sound in virtual environments can be implemented by using prerecorded sound sources. This requires sounds to be indexed, or labeled, so that they can be retrieved according to some protocol. A system based entirely on prerecorded audio files is probably the most straightforward solution, though it may be not flexible enough for some applications. Another approach is the use of fully synthesized sound by algorithms controlling a variety of parameters such as fundamental frequency, number of harmonics, intensity, and so forth. Among the most well-known algorithms for sound synthesis are additive synthesis, frequency modulation, subtractive synthesis, and granular synthesis (Cook, 2002). It may also be the case that depending on the requirements for the particular training system, both audio recordings and algorithmic generation of sound can be combined.

Independent of whichever the strategy used to generate sound, an important aspect in the design of multimodal interfaces is proper synchronization between the signals rendered for the different modalities.

AUDIOVISUAL SYNCHRONIZATION

Sensitivity to audiovisual synchronization is a useful perceptual measure to quantify the available processing time for the generation of sound. The study in Miner and Caudell (1998) measured thresholds for detecting delays between the onset of a visual signal and corresponding audio signal for single-impact stimuli, repetitive-impact stimuli, and speech. Average thresholds for impact stimuli, including single and repetitive, were 175 ms, with the lowest threshold of 100 ms. Thresholds for the audiovisual speech stimuli were higher than those for impact stimuli (200 ms). From a review on audiovisual synchronization perception (Kohlrausch and van der Par, 2005), it appears that in general humans are more sensitive to audiovisual asynchrony when audio is leading than when audio is lagging the visual signal. The fact that these thresholds are often measured with the visual stimuli presented first may have ecological reasoning. In the real world, the speed of light is one million times faster than the speed of sound. Thus in every real life situation, audio will lag.

AUDIOTACTILE SYNCHRONIZATION

In the context of asynchrony between auditory and haptic modalities, the study in Adelstein et al. (2003) measured the just-detectable delay between a voluntary hammer tap and its auditory consequence. The haptic stimulus was real, and the auditory stimulus (impact sound) was synthesized and delivered via headphones. To mask the real sound produced by the hammer tap, noise was continuously presented over the same headphones. Three different durations were employed for the auditory stimuli. The average threshold for detecting audiotactile asynchrony was 24 ms, with individual thresholds ranging from 5 to 70 ms. Duration of auditory stimuli did not have any significant effect, suggesting that listeners rely on sound onset as the primary cue. In another study (Altinsoy, 2003) that used broadband noise as auditory stimulus and a sine wave as tactile stimulus, thresholds were found in the range of –31 to 75 ms. Negative values indicate auditory stimulus preceding the haptic stimulus. Note that these studies have examined sensitivity to audiotactile asynchrony by presenting the haptic stimuli to the subjects' hand or finger. If the haptic stimuli is presented to the whole body it seems that audio delays around 10 to 20 ms are already judged as nonsimultaneous with the haptic stimulus (Martens and Woszczyk, 2004). Presumably, this increase in sensitivity is due to an increase in the effective area of tactile stimulation.

VIRTUAL SPATIAL SOUND

To properly generate virtual spatial sound one first needs to understand the perceptual cues for sound localization. The underlying psychophysical mechanisms of human sound localization are well understood (Blauert, 1996). The physical distance between our ears provides us with binaural cues derived from interaural differences in time and intensity. These cues are predominantly used for localization in the horizontal plane of a listener. In addition, sound reaching our ears is modified by the pinna, head, shoulder, and torso, which provide cues that enable us to resolve the elevation of a source as well as whether the source is in the back or in the front. All these sound localization cues are described in the head-related transfer function (HRTF).

To render virtual spatial sound, digital filters representing HRTFs are typically implemented in a process called binaural synthesis. The synthesis is performed by filtering a spatially neutral sound with a pair of HRTF filters—one filter for each ear. If done properly, the sound can be made to seem as coming from a source with a direction corresponding to that of the HRTF filter. For practical purposes spatial sound systems designed for training are often based on generic HRTFs (usually referred to as nonindividualized HRTFs). Generic HRTFs are typically obtained from artificial heads or selected from the ears of a "representative" or "typical" individual (Møller et al., 1996a). However, because generic HRTFs differ from the listener's own HRTFs, distortions to the spatial cues are introduced that may impair localization performance (Wenzel et al., 1993). Fortunately, dynamic cues derived from head movements can reduce the impact on localization of the distortions introduced by the use of generic HRTFs. This is a relevant point because head movements become particularly important in the design of interactive spatial sound systems.

Interactive or dynamic spatial sound synthesis relates to the simulation of moving sound including moving sources and compensation for listener's movement. Three engineering aspects must be considered in the implementation of such dynamic systems: latency, update rate, and spatial resolution. Latency is the time between a change in the acoustic properties of the sound field and the reflected change in the system parameters at the output. Latencies below 60 to 70 ms are likely to be adequate for most virtual reality applications (Brungart et al., 2006), although we should note that latencies as low as 30 ms can still be detected by some listeners (Sandvad, 1996). Update rate refers to the frequency at which new information is retrieved. For dynamic spatial sound this information corresponds to position and orientation of the users and sound sources. Appropriate update rates depend on the fastest velocity of moving sound that requires rendering. Nominal values of about 50 to 60 Hz appear to be adequate for most applications although faster rates are possible due to the growing increase in available computer power. It is important that the rendered sound is perceived as being updated without delays and changing smoothly, otherwise audible artifacts may be introduced that will reduce the perceptual quality of the rendering (Hoffmann and Møller, 2008). Spatial resolution is also dependent on the fastest velocity that needs rendering, and in this sense it is typically evaluated together with update rate. When sound directions cannot directly be represented by HRTF filters (e.g., due to memory constraints), interpolation is commonly used. For interpolation between HRTF filters in the time domain, it has been suggested that a resolution of 4 to 24 degrees, depending on the spatial region, is adequate so that interpolation errors are inaudible (Minnaar et al., 2005).

AUDITORY INTERFACES FOR CAPTURING

Models of skill can be derived from the analysis of multimodal information captured on experts performing the task for which the skill is defined. In the case of a sensory-motor skill, for example, capturing multimodal information includes motion capturing (i.e., experts' movements, as well as capturing the changes in the visual and auditory characteristics of the environment that are relevant for skilled performance). With regard to the auditory modality, there is particular interest in obtaining a representation of what the expert may be listening to, and thus sound capturing at the ears is essential. Binaural recordings can capture all the spatial properties of a given acoustic environment including the characteristics particular to the individual sound sources, the modifications imposed by the environment, and the anatomical filtering imposed by the listener as sound reaches the ears. In connection to the rowing training system described in Chapter 11, we discuss an auditory interface implemented for capturing binaural sound during rowing.

Qualitative descriptions of what elite rowers use as indicative of optimal rowing technique include the sounds produced by blades and boat motion (Lippens, 2005). As we mentioned before, since the sound of rowing at the ears of the athletes may be relevant for the proper representation of performance, sound capturing is done by placing miniature microphones at the entrance to the open ear canals. This technique is referred to as the open-ear canal recording technique. Figure 8.1a shows a wearable binaural recording system and illustrates its use as it is worn by a rower. The

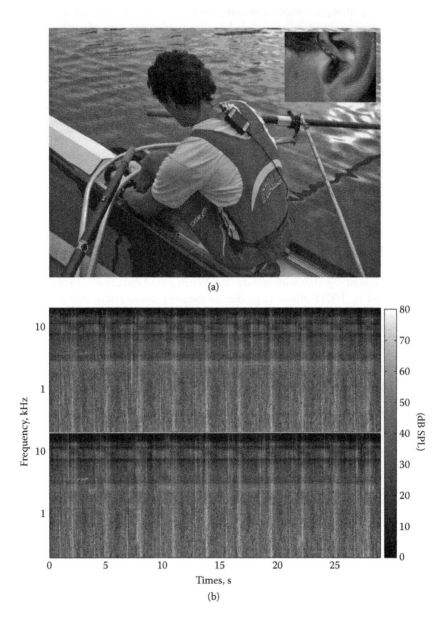

FIGURE 8.1   (a) A rower wearing the binaural recording system. The inset provides a closer look how the miniature microphone is placed at the entrance to the ear canal. (b) Spectrogram of binaural audio captured at the left (top panel) and right (bottom panel) ears for a speed of 20 strokes per minute. Gray scale represents sound pressure level (SPL) in dB re. 20μPa.

wearable binaural recorder is capable of capturing high-quality audio without disrupting the user's hearing and doing capabilities. We emphasize this property because it is critical that the recording equipment does not have a detrimental effect on the actions performed by the rower. Figure 8.1b shows a time-frequency representation, or spectrogram, of a binaural recording for a rowing speed of 20 strokes per minute (SPM). The cyclic nature of rowing is clearly represented by the periodic bursts of energy observed in the spectrogram. Roughly, the sound energy increases by about 20 to 30 dB over a frequency range of 200 to 6000 Hz. This sound event is caused by the *finish phase* of the stroke cycle. It seems that this event is a good candidate for the automatic segmentation of the rowing stroke based exclusively on acoustic information.

Ongoing work is being conducted (Johard et al., 2011) to evaluate different classification algorithms in order to determine the potentials for automatic segmentation of stroke phases, and posterior automatic classification of level of expertise based on individual analysis of the different stroke phases. It is worth noting that the main domain application is in the assessment of transfer of learning from the virtual environment to the real world.

A characteristic of open-ear canal recordings is that these recordings include the acoustics of the user's ear canal. This is generally not desired for the reproduction of these recordings on other people's ears because the acoustics of the ear canals are highly individual. An equalization procedure is necessary to compensate for this effect. A technique has been proposed in Hammershøi et al. (2008) to transform open-ear recordings to blocked-ear versions because blocked-ear recordings can capture all the acoustic attributes of the recorded sound field but without the highly individual influence of the acoustics of the ear canal.

AUDITORY INTERFACES FOR RENDERING

There are many ways of rendering sound in virtual environments (Kapralos et al., 2008). Here we first discuss the use of binaural reproduction as part of the sound rendering system implemented for the rowing scenario described in Chapter 11. Next we focus our discussion on a different form of rendering; namely, we describe the analysis-synthesis method used to implement the drilling sound in the surgical simulator of the maxillofacial surgery described in Chapter 13 of this book.

ROWING TRAINING SYSTEM

The system for sound rendering in the rowing simulator (see Chapter 11) has been implemented with the aim of reproducing real-life sounds recorded at the ears of expert rowers as described above. The rendering technique is based on concatenative synthesis in the sense that segments of binaural recordings are selected according to a given criterion, concatenated, and reproduced as a continuous binaural stream. For each velocity, given in SPM, and stroke phase, three sound segments were manually selected from the binaural recordings of the most experienced rower. A database of audio files was built using these sound segments. Audio files were stored in WAV format (16-bit and 48-kHz sampling rate) and had a fixed duration of 2 seconds. The duration of the longer stroke phase was approximately about 1.2 seconds (18 SPM),

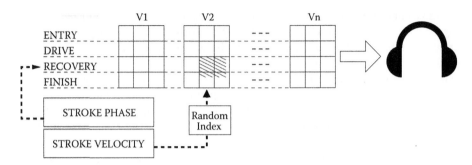

FIGURE 8.2   The sound rendering system for the rowing simulator. Audio files are described by boxes grouped by velocity (V1, V2, ..., Vn) and phase. Information on stroke phase and velocity is received at a 100-Hz rate, and it is used to select the appropriate sound (indicated as an example by the shaded box). For every combination of phase and velocity there are three possible sound samples.

and thus each audio file in the database started with the sound of its corresponding phase, and the remaining part was filled with environmental sound captured during moments when the rower was not rowing and the boat was not in motion. Figure 8.2 provides a schematic description of the rendering system. The inputs to the sound renderer were stroke velocity and phase that were calculated according to the angles of the oars in the simulator (see Chapter 11 for further detail). Stroke phase could take the value of four labels corresponding to 1, entry; 2, drive; 3, finish; and 4, recovery.

The rowing simulator runs with an update rate of 100 Hz—that is, every 10 ms the system sends velocity and phase information to the sound module that automatically selects at random one of the three audio files corresponding to that particular combination of velocity and phase (as an example, the selected sound is depicted by the shaded box in Figure 8.2). Having more than one sound per velocity and phase makes the sound rendering less predictable.

Maxillofacial Surgery Training System

There is a long history of virtual environment training systems for medical applications. In the surgical domain, the great benefit of virtual training is that trainees can practice in a risk-free environment where mistakes can be made without compromising the health of real patients. Typically, training simulators make extensive use of visual and haptic interfaces to provide the desired simulation fidelity for the targeted surgical procedure (Liu et al., 2003). Not so much attention has been given to the auditory modality, and this is probably because many simulators target surgical procedures in which sound may not play a critical role. However, surgical procedures involving drilling are expected to have a significant auditory component.

Here we are concerned with the integration of an auditory interface in a virtual training system designed to train a maxillofacial surgery. See Chapter 13 for a detailed description of this training system.

The main reason for including an auditory interface in this training system is that, besides the haptic cues normally available in surgical drilling procedures, there is evidence that auditory cues may also play a relevant role in expert drilling performance. The study by Praamsma et al. (2008) reported an experiment that evaluated whether orthopedic drilling skills depended on the availability of auditory information produced by drilling, and whether this dependency was in turn dependent on level of expertise. Experienced surgeons, intermediate residents, and novices had to perform a drilling procedure in two conditions distinguished by the absence or presence of a broadband noise that masked the drilling sound. Their results showed that expert surgeons' performance and residents' performance were significantly better than novices' performance, and more importantly, that expert and residents' performances were significantly affected by the presence of noise, but novices' performance remained unaffected. Although it is possible that expert surgeons and residents were simply more annoyed by the noise than novices, it seems more plausible that novices could not make effective use of auditory cues as a source of critical information for more accurate drilling. This suggests that training protocols may need to carefully consider the proper elaboration of guidelines that facilitate learning of the auditory cues available during surgical drilling.

Proper simulation of the maxillofacial surgery requires providing the relevant haptic and auditory feedback in response to the forces applied during drilling. A strategy for the reproduction of drilling sounds similar to that of the rowing simulator would require many short recordings in order to represent a suitable range of forces and various stages of burr-tissue interactions. Considering the many possible combinations of burr-tissue interactions and forces, this strategy is not feasible. A more viable solution is to use sound synthesis in which the sound is represented by a parametric model.

A sinusoidal model has been used to synthesize the sound produced by the drilling procedure in the maxillofacial surgery simulator (Hoffmann, Gosselin, and Taha, 2009; Hoffmann et al., 2009). The parameters of a sinusoidal model are frequency, amplitude, and phase. Audio recordings were obtained from a multimodal capturing session conducted while expert surgeons perform a complete maxillofacial surgery on cadavers. The sinusoidal model extracts the parameters of a sinusoidal component from windowed segments of the original sound recordings using a matching pursuit algorithm (Heusdens et al., 2002; Verma and Meng, 1999). An important engineering question concerns the minimum number of components required to synthesize a drilling sound so that the synthesized version is perceptually equivalent to the original sound. Perceptual equivalence can be estimated by measuring the just noticeable difference (JND) between the original and synthesized drilling sound. Since the duration of the analysis window also has an impact on the quality of the synthesized sound, a listening experiment was conducted to assess the discrimination of synthesized sounds from the real sounds for a discrete combination of number of sinusoidal components and window duration.

Four subjects with normal hearing participated in this experiment. All subjects were experienced in psychoacoustic experiments. Sixteen drilling-sound excerpts of 1 second duration were manually selected from an audio recording of a maxillofacial surgery performed by an expert surgeon. The analysis window was a Hanning window and the durations used were 5.3, 10.7, and 21.3 ms with 50% overlap between consecutive frames. For each of the different window durations sounds were synthesized

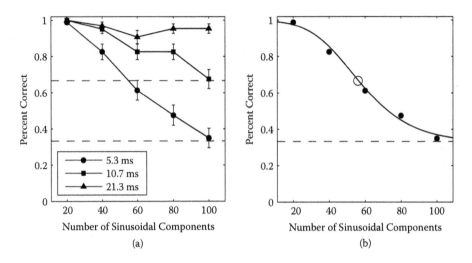

FIGURE 8.3 (a) Proportions of correct discrimination between original and synthesized drilling sounds. Different symbols indicate different window durations. Error bars indicate ±1 standard deviation. (b) Psychometric function fitted to the proportions corresponding to 5.3 ms of duration. Estimated threshold indicated by the open circle.

using 20, 40, 60, 80, and 100 sinusoidal components that were extracted from the original sound excerpts using the sinusoidal model mentioned above. Discrimination of differences between original and synthesized sounds was estimated in a 3-interval 3-alternative forced-choice (3AFC) procedure using the method of constant stimuli. To avoid the use of intensity cues for discrimination, original and synthesized sounds were equated on their root mean square (RMS) values. Stimuli were delivered to the listener over equalized Beyerdynamic DT-990 circumaural headphones.

Figure 8.3 (left panel) shows the average proportion of correct responses as a function of number of sinusoidal components and window durations. For 20 sinusoidal components listeners could fully discriminate sounds independent of window duration. As the number of sinusoidal components increased, the difference in performance across window durations increased. For 21.3-ms duration, performance remained fairly constant across number of sinusoidal components. For durations 10.7 and 5.3 ms performance decreased with number of sinusoidal components, with a steeper slope for the shortest duration. The lower dashed line in the left panel of Figure 8.3 indicates the theoretical random-response level corresponding to the reciprocal of the number of alternatives in the forced-choice procedure (i.e., 1/3 or 33%). By defining the point of perceptual equality as halfway between random performance and perfect performance, our point of perceptual equality corresponded to 2/3 or 66% correct response and is indicated by the upper dashed line.

In order to estimate a threshold value, a psychometric curve was fitted to the proportions associated to the window length of 5.3 ms using the procedure described in Zychaluk and Foster (2009). This result is shown in the right panel of Figure 8.3. For this psychometric function the estimated number of sinusoidal components required to generate a 66% correct response is 56, thus we decided that a total of 60 components would be a safe choice for the synthesis of drilling sound.

As described in Hoffmann, Gosselin, and Taha (2009) and Hoffmann et al. (2009), an additional processing unit can be applied to the synthesized drilling sound in order to further reduce the number of sinusoidal components. This processing unit makes use of a simple model of simultaneous masking in the auditory system (Moore, 2003). From the computed 60 components on a given drilling-sound segment, a masking curve is constructed using the concept of excitation pattern (van der Heijden and Kohlrausch, 1994). An excitation pattern can be roughly described as the result from the spectral analysis done in the cochlea (it is analogous to the pattern of vibration on the basilar membrane). Once the masking curve is computed it is compared against the amplitudes of the sinusoidal components, and all components whose amplitudes are below the masking curve are discarded. Generally it is possible to achieve an average reduction of 50%. That is, a perceptually adequate synthesis of the drilling sound in maxillo-facial surgery can be based on approximately 30 sinusoidal components. To compare the quality of the drilling sound with 60 components against that of the psychoacoustically based synthesis, a standardized method known as PEAQ (Perceptual Evaluation of Audio Quality) was used (ITU-R Rec. BS.1387 1998). The method computes the objective difference grade (ODG) that classifies the perceptual quality of sounds relative to the original high-quality sound based on the following scale: 0, imperceptible; –1, perceptible but not annoying; –2, slightly annoying; –3, annoying; –4, very annoying.

Figure 8.4 shows the comparison between the ODG scores obtained for the synthesis using 60 sinusoidal components and the synthesis using the reduced number of components based on the psychoacoustic masking model. What we emphasize here is that since the 60 sinusoidal components can produce a synthesized drilling sound that is not discriminable from the original sound, and that the ODG difference

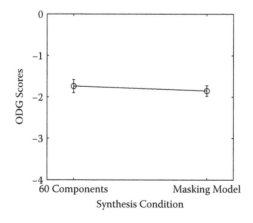

FIGURE 8.4 Objective difference grade (ODG) scores for drilling sounds synthesized using 60 sinusoidal components and the psychoacoustic masking model with a reduced number of components.

between the 60 sinusoidal components sound and the masking model sound is not significant, then the sound synthesized using the masking model (reduced number of components) is also perceptually equivalent to the original sound.

AUDITORY INTERFACES IN MULTIMODAL PERCEPTION OF SELECTED SKILLS

The spatial component of the auditory modality provides useful features for the design of effective auditory interfaces targeting skill training and transfer. Some of these features include simultaneous hearing of sounds spatially arrayed, faster detection than signals presented through the visual channel, and tendency to produce an alerting and orienting response (Ellis et al., 1998).

Clearly, the most intuitive application of spatial sound for training is in the improvement of skills associated to spatial perception. In Jones, Stanney, and Foaud (2005) an experiment was reported that measured performance in a military virtual training operation. Participants had to move around a virtual environment and engage hostile and nonhostile units. The environment consisted of several rooms, hallways, doors, and windows. Sounds of weapon fires, voices, and movements of hostile and nonhostile units were spatialized using HRTFs. Four conditions were compared: a control condition of no sound, audio without spatial information, spatial sound using generic HRTFs, and spatial sound using a procedure to determine best-fit HRTFs according to a custom sound-localization criterion. Using time for task completion as a performance measure, it was found that only after a period of training the integration of spatial sound significantly reduced the time required to complete the task. There were no differences between the use of generic HRTFs and the use of best-fit HRTFs.

SOUND LOCALIZATION SKILLS AND SENSORY-MOTOR LEARNING

Given that determining the position of a sound source in space is an integral part of how humans interact with multimodal environments, the importance of adequate sound localization in complex auditory environments promotes examination of the usability of feedback from other modalities to improve performance on sound localization tasks. For example, it has been shown that physically based visual information facilitates auditory localization judgements (Shelton and Searle, 1980). The study in Majdak, Goupell, and Laback (2010) reported an experiment that used a virtual visual environment to examine the differential effect of providing only proprioceptive cues and both proprioceptive and visual cues on localization performance of sound spatialized using individual HRTFs. Their findings suggest that visual and proprioceptive information provided by the virtual environment can significantly improve sound localization performance relative to only providing proprioceptive cues, and that the response method, either head-pointing or manual pointing, did not have a significant influence.

Further, a number of studies have provided evidence that auditory localization ability in human adults improves with training, and that training-induced learning occurs with both natural spatial cues and altered cues (Wright and Zhang, 2006). Learning with natural cues demonstrates the capacity of the auditory system to fine-tune itself through experience. Learning with altered cues indicates the capacity of the auditory system to assimilate and comprehend new spatial relations between source position and its percept. An interesting aspect here is that aftereffects of training with unnatural cues appear to be temporary and small in magnitude. That is, learning a new set of unnatural spatial cues does not seem to disrupt localization performance with natural cues (Hofman et al., 1998). As we will discuss further below, this aspect may prove to be a relevant factor in the design of spatial auditory interfaces for virtual environment training when the interfaces are based on generic representations of auditory space (e.g., nonindividual HRTFs).

A direct reflection of the notion that we learn to listen through experience with our own ears is evidenced by the well-established knowledge indicating that we localize best when we listen to binaural signals that have been recorded at our own ears (Møller et al., 1996b). Another indication of this phenomenon, although less direct, is given by the significant decrease in localization performance of virtual sounds synthesized using nonindividual HRTFs as compared with localization of real sources (Wenzel et al., 1993). This result would imply that in order to build a usable spatial auditory interface, one would need to measure the HRTFs of every potential user of such an interface, clearly making its implementation impractical. The study in Zahorik et al. (2006) presented results indicating that spatial perception of sound in the front and behind can be learned, as reflected by a significant reduction in front-back reversals in the trained group, when using a multisensory training protocol based on auditory, visual, and proprioceptive, or vestibular, feedback. An interesting aspect of this study is that the filters used to simulate spatial sound were based on nonindividual HRTFs, and that learning occurred within a relatively short training time of two 30-minute sessions separated by at least 1 day in a time span of 4 days. This learning time corresponded to a total of 144 trials. Learning of spectral cues for elevation perception has been shown to require longer training periods (Hofman, Van Riswick, and Van Opstal 1998), and this could be attributed to the idea that the cues used to resolve front-back confusions are only a subset of the cues used for elevation perception. The results from Zahorik et al. (2006) have interesting practical implications. That is, learning to effectively use nonindividual HRTFs in a relatively short time suggests that instead of adapting the auditory interface to the user, as it would have been the case if measuring HRTFs for every potential user was required, the user can adapt to the interface via training. This is further supported by additional results in Zahorik et al. (2006) showing that participants retained the learned localization performance for a period of about 3 months.

Another interesting aspect in the usability of auditory interfaces for virtual environment training is the engaging or motivating role that entertainment may play on the training power of the auditory interface. In the study by Honda et al. (2007) subjects were trained on sound source localization by playing an interactive auditory spatial game. It can be argued that a gaming metaphor for training has the potential to enhance motivation, which in turn can facilitate auditory learning (Amitay et al., 2010). The game in Honda et al. (2007) consisted of hitting virtual bees that could be localized exclusively through sound. The sound of a bee was spatialized using individualized HRTFs for one training group and nonindividualized HRTFs, derived from a fitting procedure, for a second training group. Localization of real sound sources was measured on pre- and post-training tests. A third control group also participated in the pre- and post-training tests but did not go through the training phase. During training, subjects were instructed to play the game for 30 minutes per day during 7 days on a period of 2 weeks. Results indicated that both training groups significantly improved sound localization performance relative to the control group. Similar to the retention results presented by Zahorik et al. (2006), the trained participants showed that the transfer effect on localization performance remained after a month.

An important factor in the efficacy of auditory interfaces for virtual environment training is how well the population of targeted trainees can adapt to the interface so that they can make an effective use of the interface. In addition to the study by Zahorik et al. (2006) that showed that 144 trials appear to be enough for improving sound localization accuracy via reduction of front-back reversals, results from a second experiment in Majdak, Goupell, and Laback (2010) designed to evaluate sound-localization training in a visual virtual environment suggested that within the first 400 trials the largest improvement in sound localization performance is observed. After the first 400 trials, improvements were small and the corresponding rate of learning slowed down but it did not reach asymptotic performance even after 1600 trials. A possible explanation for the difference in the number of trials required to achieve learning in these two studies may be because the 144 trials seem to consider improvements in front-back reversals only, whereas 400 trials seem to include reductions in azimuth and elevation errors in addition to reductions in front-back reversals. Interestingly, in the examination of the effects that practice has on a visual-target acquisition task aided by spatial auditory cues based on nonindividual HRTFs, results reported in McIntire et al. (2010) also suggest that effective use of spatial auditory cues appears to be attained after a training period of about 300 trials.

Taken together, it appears that visual and proprioceptive feedback provided as part of a virtual training environment is effective at improving accuracy and performance on auditory localization, independent of whether spatial auditory interfaces are based on individual or nonindividual HRTFs. In addition, it appears that effective use of these interfaces can be attained after a relatively short training time, which, from a practical viewpoint makes their use attractive. Although further research is clearly required, some evidence suggests that comparable practice times are observed for training sound localization in virtual environment and for training localization tasks engaging other modalities (e.g., visual-target acquisition in which sound localization is used to enhance task performance).

AUDITORY SPATIAL INTERFACES FOR THE VISUALLY IMPAIRED

When deprived of one sensory modality individuals tend to compensate its absence by improving, or specializing, the use of the remaining functional modalities. For blind individuals, for example, the work reported in Merabet and Sánchez (2009) used a game paradigm in an interactive virtual auditory environment to aid blind users to make effective use of auditory spatial cues in a navigation task. The benefits of auditory cues for navigations were evaluated by asking the participants to reconstruct the virtual space they navigated by building a Lego® model. Their results suggest that blind individuals could generate accurate mental representations of the auditory space, implying a high degree of fidelity in the spatial cognitive maps generated by the gaming paradigm of the experiment. The use of interactive, immersive gaming metaphors in training seems to provide a powerful and motivating drive. Again, this is particularly relevant considering that motivation can modulate auditory learning (Amitay et al., 2010). In addition to auditory orientation training, training on auditory obstacle perception and avoidance is a sensory-motor skill critical for the successful realization of daily activities by visually impaired individuals. An auditory orientation training system has been recently developed (Seki and Sato, 2011) that includes obstacle perception via the simulation of reflections from surfaces. The system is an interactive three-dimensional sound system based on HRTFs. Four elements can be rendered in the training environment, namely, sound sources (e.g., vehicles, pedestrians), walls, roads, and landmarks. Results from a test using blindfolded subjects suggested that the system was the most effective method to train auditory orientation skills relative to a control group that did not train, and a group trained using a standard orientation training protocol in a real environment.

Auditory Facilitation of Visual Tasks

The auditory spatial channel plays a significant role in regulating visual gaze, indicating that one of the primary functions of spatial auditory processing is to guide visual orienting (Perrott et al., 1990). Incorporating spatial audio interfaces to cue the position of a target in a visual search task can significantly reduce target acquisition times relative to an uncued condition. Studies have shown that performance improvement occurs for static visual targets (Bolia et al., 1999; Perrott et al. 1996), and that this improvement transfers from static targets to dynamic visual targets (McIntire et al. 2010). Besides, the enhancing effect of localized auditory information is more pronounced as the complexity of the task or visual load increases (Perrott et al., 1991). Supplementing visual search with information resulting from including localized auditory cues does not compromise performance in the accuracy of identifying the target (Bolia et al., 1999). Also, it should be emphasized that visual search times are consistently faster with spatially informative auditory cues than with uninformative cues (Flanagan et al., 1998; Nelson et al., 1998; Rudmann and Strybel, 1999; Vu, Strybel, and Proctor, 2006), thus suggesting that it is the actual spatial attribute of the cue that benefits performance in visual search and not the mere presentation of an auditory alert. The studies reported in Rudmann and Strybel (1999) and Vu, Strybel, and Proctor (2006) have also shown that performance is hindered when the auditory cues are displaced around the visual target, thus suggesting a difficulty in ignoring auditory cues even when they supply incorrect information.

Spatial auditory interfaces have been shown to provide positive effects in terms of increasing situational awareness in high-stress applications (e.g., aircraft cockpits), and also to increase engagement by improving the sense of immersion in training virtual environments. Therefore, it is not surprising to find spatial sound technology incorporated in the design of multimodal interfaces for military training applications. The study in Bronkhorst, Veltman, and van Breda (1996) investigated the utility of a spatial auditory display in a flight simulation experiment. Pilots had to localize a target using only visual cues, only auditory cues, both visual and auditory cues, and no cues in either modality. Visual and auditory cues were equally effective in providing information for target acquisition. When cues from the two modalities were presented simultaneously, highest performance was reached. The study in Oving, Veltman, and Bronkhorst (2004) showed that using spatial auditory displays can improve about 12% reaction times to warnings in a Traffic Alert and Collision Avoidance System (TCAS) found in most commercial aircrafts. It was also observed that the advantage of spatial sound was particularly strong when the visual channel was focused on other tasks.

In Parker et al. (2004) participants were required to perform a visual-target acquisition task while following a leading aircraft. Spatial sound was simulated using listeners' own HRTFs. To increase the workload, on each trial three to four distracters were presented along with the target. There was a significant reduction of about 30% in mean visual acquisition time for spatial sound (from approximately 14 s to 10 s). Time used to scan the visual display during target presentation was significantly larger for the visual condition than for the condition in which spatial sound was integrated with vision (500 to 600 ms of reduction). In addition, when spatial auditory information was available, participants referred to the visual display around 22% of trials in contrast to the 90% to 92% observed when only visual information was available. The authors argued that listeners' little experience in using visual displays may have been a factor in the large effect of visual-display scanning between nonspatial and spatial sound conditions. Situational awareness increased with spatial sound as revealed by a subjective test.

Using a simple assembly task that consisted of fitting a peg into a hole made on a plate, the study in Zhang, Xiao, and Travis (2006) examined the effects of auditory, visual, and audiovisual feedback on performance, and these were compared to performance on a baseline condition corresponding to no feedback. Auditory stimulus was delivered over headphones and was presented with full spatial information. Objective measures of performance were time to task completion and number of failures. A significant improvement was observed when auditory feedback operated relative to the baseline condition. In a more complex task that consisted of the assembly of an electronic box, preference and helpfulness were significantly better when auditory feedback was provided. In general, either visual or auditory feedback showed a positive effect on task performance and subjective preference. When auditory and visual feedbacks were integrated, results showed the best performance and highest preference ratings.

AUDITORY FACILITATION OF TACTILE TASKS

Evidence that auditory stimuli can modulate tactile perception can be an important aspect for the skill *perception by touch*. It has been shown that people can discriminate tactile roughness on the basis of auditory feedback alone, but when combined audiotactile information is available tactile cues are weighted more heavily than auditory cues (Lederman, 1979). It has been hypothesized that tactile roughness perception can be modulated by frequency content of sound as well as intensity (loudness). In Guest et al. (2002) results from one of the experiments show that amplifying high frequencies (12 dB boost of 2 to 20 kHz) increases roughness perception, and that attenuating the same frequency range (–12 dB) leads to an increased perception of smoothness. In general, it appears that manipulating the sounds associated with touching a surface can modulate people's perception of the texture of the surface they are touching. As pointed out by the authors, an interesting study could be to investigate whether people modulate the force with which they touch a surface in response to the loudness of the auditory feedback they hear. In addition, DiFranco, Beauregard, and Srinivasan (1997), Avanzini and Crosato (2006a) provide evidence that, besides roughness perception, auditory feedback can modulate perception of stiffness when using a rigid probe.

In connection to industrial applications, the study in Edwards and Barfield (2004) evaluated the effectiveness of auditory feedback as substitute of force feedback in an assembly task. Time to complete the task and number of collision were used as objective measures of performance. Auditory feedback was provided via headphones for the effects of collision (metallic sound), grasping, and assembling (a clicking sound indicated correct assembling). For completion time, it was found that auditory feedback did not significantly improve performance as compared to the baseline condition (no feedback in either modality) and force feedback condition. Auditory feedback showed a reduction in number of collisions as compared to the force feedback condition, although the reduction was modest. Subjective evaluation of the virtual reality system revealed that auditory feedback improved participants' perception of realism and satisfaction. In this study, auditory spatial information was not incorporated. In a similar study conducted by Petzold et al. (2004), results did not indicate that auditory feedback could be an effective substitution of force feedback. It was found that auditory feedback could enhance the sense of presence, but only to a modest extent. Auditory feedback was given in the form of a metallic sound indicating collision events.

CONCLUDING REMARKS

Virtual environments and multimodal interfaces open the possibility to enhance human performance on several skills and tasks or application contexts. Further research is required for a more detailed assessment of the effect that different modalities may have on skill acquisition, and to determine which modality may be more relevant for specific tasks. The research summarized in this chapter indicates that in training mediated by multimodal virtual environment technologies, the auditory modality can play a significant role in the effective acquisition of skills involving temporal and spatial processing.

ACKNOWLEDGMENTS

The work presented in this chapter was supported by the SKILLS integrated Project (IST-FP6 #035005, http://www.skills-ip.eu) funded by the European Commission.

The invaluable help of Søren Krarup Olesen for his scientific and technical contributions, and Claus Vestergaard Skipper and Peter Dissing for all their technical support throughout the life span of SKILLS is sincerely acknowledged.

REFERENCES

Adelstein, B.D., D.R. Begault, M.R. Anderson, and E.M. Wenzel. 2003. Sensitivity to haptic-audio asynchrony. Paper presented at ICMI '03: *Proceedings of the Fifth International Conference on Multimodal Interfaces*, Vancouver, British Columbia, Canada.

Ahmad, A.M., K.M. Stanney, and H. Fouad. 2009. Theoretical foundations for integrating sound in interactive interfaces: Identifying temporal and spatial information conveyance principles. *Theoretical Issues in Ergonomics Science* 10 (2): 161–186.

Altinsoy, M.E. 2003. Perceptual aspects of auditory-tactile asynchrony. Paper presented at *Proceedings of the 10th International Congress of Sound and Vibration*, Stockholm, Sweden.

Amitay, S., L. Halliday, J. Taylor, E. Sohoglu, and D.R. Moore. 2010. Motivation and intelligence drive auditory perceptual learning. *PloS One* 5 (3): e9816.

Avanzini, F., and P. Crosato. 2006a. Haptic-auditory rendering and perception of contact stiffness. Paper presented at *Proceedings of the First International Workshop on Haptic and Audio Interaction Design (HAID)*, Glasgow, UK.

Avanzini, F., and P. Crosato. 2006b. Integrating physically based sound models in a multimodal rendering architecture. *Computer Animation and Virtual Worlds* 17: 411–419.

Blauert, J. 1996. *Spatial Hearing: The Psychophysics of Human Sound Localization*. Revised Edition ed. Cambridge, MA: MIT Press.

Bolia, R.S., W.R. D'Angelo, and R.L. McKinley. 1999. Aurally aided visual search in three-dimensional space. *Human Factors* 41 (4): 664–669.

Bronkhorst, W.W., J.A.H. Veltman, and L. van Breda. 1996. Application of a three-dimensional auditory display in a flight task. *Human Factors* 38 (1) (March): 23–33.

Brungart, D.S., A.J. Kordik, and B.D. Simpson. 2006. Effects of headtracker latency in virtual audio displays. *Journal of the Audio Engineering Society* 54 (1/2) (January): 32–44.

Cook, P. 2002. *Real Sound Synthesis for Interactive Applications*. Natick, MA: A.K. Peters.

DiFranco, D.E., G.L. Beauregard, and M.A. Srinivasan. 1997. The effect of auditory cues on the haptic perception of stiffness in virtual environments. Paper presented at *Proceedings of the ASME, Dynamic Systems and Control Division*.

Edwards, G.W., and W. Barfield. 2004. The use of force feedback and auditory cues for performance of an assembly task in an immersive environment. *Virtual Reality* 7 (2): 112–119.

Ellis, S.R., D. Begault, and E.M. Wenzel. 1998. Virtual environments as human-computer interfaces. In *Handbook of Human-Computer Interaction.*, eds. M. Helander, T.K. Landauer, and P.V. Prabhu. North-Holland: Elsevier.

Flanagan, P., K.L. McAnally, R.L. Martin, J.W. Meehan, and S.R. Oldfield. 1998. Aurally and visually guided visual search in a virtual environment. *Human Factors: Journal of the Human Factors and Ergonomics Society* 40 (3): 461–468.

Garcia-Ruiz, M.A., and J.R. Gutierrez-Pulido. 2006. An overview of auditory display to assist comprehension of molecular information. *Interacting with Computers* 18: 853–868.

Gonot, A., N. Chateau, and M. Emerit. 2006. Usability of 3D-sound for navigation in a constrained virtual environment. Paper presented at *Proceedings of the 120th Convention of the Audio Engineering Society*, Paris, France.

Guest, S., C. Catmur, D. Lloyd, and C. Spence. 2002. Audiotactile interactions in roughness perception. *Experimental Brain Research* 146: 161–171.

Hale, K.S., K.M. Stanney, L.M. Milham, M.A.B. Carroll, and D.L. Jones. 2009. Multimodal sensory information requirements for enhancing situation awareness and training effectiveness. *Theoretical Issues in Ergonomic* 10 (3): 245–266.

Hammershøi, D., P.F. Hoffmann, S.K. Olesen, and P. Rubak. 2008. Capturing blocked-entrance binaural signals from open-entrance recordings. Paper presented at *Proceedings of Acoustics '08 Société Francaise d'Acoustique*, Paris, France.

Heusdens, R., R. Vafin, and W.B. Kleijn. 2002. Sinusoidal modeling using psychoacoustic-adaptive matching pursuits. *IEEE Signal Processing Letters* 9 (8) (August): 262–265.

Hoffmann, P.F., and H. Møller. 2008. Audibility of direct switching between head-related transfer functions. *Acta Acustica United with Acustica* 94 (6): 955–964.

Hoffmann, P.F., F. Gosselin, and F. Taha. 2009. Analysis of the drilling sound component from expert performance in a maxillo-facial surgery. Paper presented at *Proceedings of the 15th International Conference on Auditory Displays (ICAD 2009)*, Copenhagen, Denmark.

Hoffmann, P.F., F. Gosselin, F. Taha, S. Bouchigny, and D. Hammershøi. 2009. Analysis of the drilling sound in maxillo-facial surgery. Paper presented at *Proceedings of the International Conference on Multimodal Interfaces for Skills Transfer*, Bilbao, Spain.

Hofman, P.M., J.G. Van Riswick, and A.J. Van Opstal. 1998. Relearning sound localization with new ears. *Nature Neuroscience* 1: 417–421.

Honda, A., H. Shibata, J. Gyoba, K. Saito, Y. Iwaya, and Y. Suzuki. 2007. Transfer effects on sound localization performances from playing a virtual three-dimensional auditory game. *Applied Acoustics* 68 (8) (August): 885–896.

ITU-R Rec. BS.1387. 1998. *Method for objective measurement of perceived audio quality, ITU, Geneva, Switzerland, 1998, ITU-R rec. BS.1387*.

Johard, L., E. Ruffaldi, P. Hoffmann, and A. Filippeschi. 2011. Machine learning analysis of binaural rowing sounds. BIO Web of Conferences 1: 43, p. 1–43. p. 4.

Jones, D.L., K.M. Stanney, and H. Foaud. 2005. An optimized spatial audio system for virtual training simulations: Design and evaluation. Paper presented at *Proceedings of the 11th International Conference on Auditory Display*, Limerick, Ireland.

Kapralos, B., M.R. Jenkin, and E. Milios. 2008. Virtual audio systems. *Presence: Teleoperators and Virtual Environments* 17 (6) (October): 527–549.

Kohlrausch, A., and S. van der Par. 2005. Audio-visual interaction in the context of multimedia applications. In *Communication Acoustics.*, ed. Jens Blauert, 109–138. Berlin, Germany: Springer Verlag.

Lederman, S.J. 1979. Auditory texture perception. *Perception* 8 (1): 93–103.

Lippens, V. 2005. Inside the rower's mind. In *Rowing Faster*, ed. V. Nolte, 185–194. Champaign, IL: Human Kinetics.

Liu, A., F. Tendick, K. Cleary, and C. Kaufmann. 2003. A survey of surgical simulation: Applications, technology, and education. *Presence: Teleoperators and Virtual Environments* 12 (6): 599–614.

Lokki, T., and M. Gröhn. 2005. Navigation with auditory cues in a virtual environment. *Multimedia, IEEE* 12 (2): 80–86.

Majdak, P., M.J. Goupell, and B. Laback. 2010. 3-D localization of virtual sound sources: Effects of visual environment, pointing method, and training. *Attention, Perception and Psychophysics* 72 (2): 454–469.

Martens, W.L., and W. Woszczyk. 2004. Perceived synchrony in a bimodal display: Optimal intermodal delay for coordinated auditory and haptic reproduction. Paper presented at *Proceedings of the 10th International Conference on Auditory Display*, Sydney, Australia.

McIntire, J.P., P.R. Havig, S.N.J. Watamaniuk, and R.H. Gilkey. 2010. Visual search performance with 3-D auditory cues: Effects of motion, target location, and practice. *Human Factors: Journal of the Human Factors and Ergonomics Society* 52 (1): 41–53.

Merabet, L.B., and J. Sánchez. 2009. Audio-based navigation using virtual environments: Combining technology and neuroscience. *Research and Practice in Visual Impairment and Practice* 2 (3): 128–137.

Miner, N., and T. Caudell. 1998. Computational requirements and synchronization issues for virtual acoustic displays. *Presence: Teleoperators and Virtual Environments* 7 (4): 396–409.

Minnaar, P., J. Plogsties, and F. Christensen. 2005. Directional resolution of head-related transfer functions required in binaural synthesis. *Journal of the Audio Engineering Society* 53 (10) (October): 919–929.

Møller, H., C.B. Jensen, D. Hammershøi, and M.F. Sørensen. 1996a. Using a typical human subject for binaural recording. Paper presented at *Proceedings of the 100th Convention of the Audio Engineering Society*, Copenhagen, Denmark.

Møller, H., M.F. Sørensen, C.B. Jensen, and D. Hammershøi. 1996b. Binaural technique: Do we need individual recordings? *Journal of the Audio Engineering Society* 44 (6): 451–469.

Moore, B.C.J. 2003. *An Introduction to the Psychology of Hearing,* 5th ed. London: Academic Press.

Müller-Tomfelde, C. 2004. Interaction sound feedback in a haptic virtual environment to improve motor skill acquisition. Paper presented at *Proceedings of the 10th International Conference on Auditory Display*, Sydney, Australia.

Nelson, W.T., L.J. Hettinger, J.A. Cunningham, B.J. Brickman, M.W. Haas, and R.L. McKinley. 1998. Effects of localized auditory information on visual target detection performance using a helmet-mounted display. *Human Factors* 40 (3): 452–460.

Novo, P. 2005. Auditory virtual environments. In *Communication Acoustics*, ed. J.J. Blauert, 277–297. Berlin, Germany: Springer Verlag.

Oving, A.B., J.A. Veltman, and A.W. Bronkhorst. 2004. Effectiveness of 3-D audio for warnings in the cockpit. *International Journal of Aviation Psychology* 14 (3): 257–276.

Parker, S.P.A., S.E. Smith, K.L. Stephan, R.L. Martin, and K.I. McAnally. 2004. Effects of supplementing head-down displays with 3-D audio during visual target acquisition. *International Journal of Aviation Psychology* 14 (3): 277–295.

Perrott, D.R., J. Cisneros, R.L. McKinley, and W.R. D'Angelo. 1996. Aurally aided visual search under virtual and free-field listening conditions. *Human Factors* 38 (4): 702–715.

Perrott, D.R., K. Saberi, K. Brown, and T.Z. Strybel. 1990. Auditory psychomotor coordination and visual search performance. *Perception and Psychophysics* 48 (3): 214–226.

Perrott, D.R., T. Sadralodabai, K. Saberi, and T. Strybel. 1991. Aurally aided visual search in the central visual field: Effects of visual load and visual enhancement of the target. *Human Factors* 33 (4): 389–400.

Petzold, B., M.F. Zaeh, B. Faerber, B. Deml, H. Egermeier, J. Schilp, and S. Clarke. 2004. A study on visual, auditory, and haptic feedback for assembly tasks. *Presence: Teleoperators and Virtual Environments* 13 (1): 16–21.

Praamsma, M., H. Carnahan, D. Backstein, C.J.H. Veillette, C. Gonzalez, and A. Dubrowski. 2008. Drilling sounds are used by surgeons and intermediate residents, but not novice orthopedic trainees, to guide drilling motions. *Can J Surg* 51 (6): 442–446.

Rudmann, D.S., and T.Z. Strybel. 1999. Auditory spatial facilitation of visual search performance: Effect of cue precision and distractor density. *Human Factors: Journal of the Human Factors and Ergonomics Society* 41 (1): 146–160.

Sandvad, J. 1996. Dynamic aspects of auditory virtual environments. Paper presented at *100th Convention of the Audio Engineering Society*, Copenhagen, Denmark.

Seki, Y., and T. Sato. 2011. A training system of orientation and mobility for blind people using acoustic virtual reality. *IEEE Transactions on Neural Systems and Rehabilitation Engineering* 19 (1): 95–104.

Shams, L., and A.R. Seitz. 2008. Benefits of multisensory learning. *Trends in Cognitive Sciences* 12 (11): 411–417.

Shelton, B.R., and C.L. Searle. 1980. The influence of vision on the absolute identification of sound-source position. *Perception and Psychophysics* 28 (6): 589–596.

Sigrist, R., J. Schellenberg, G. Rauter, S. Broggi, R. Riener, and P. Wolf. 2011. Visual and auditory augmented concurrent feedback in a complex motor task. *Presence: Teleoperators and Virtual Environments* 20 (1): 15–32.

van der Heijden, M., and A. Kohlrausch. 1994. Using an excitation-pattern model to predict auditory masking. *Hearing Research* 80 (1): 38–52.

Verma, T.S., and T.H.Y. Meng. 1999. Sinusoidal modeling using frame-based perceptually weighted matching pursuits. Paper presented at *Proceedings of the IEEE International Conference on Acoustics, Speech, and Signal Processing*, Phoenix, AZ.

Vu, K.-P.L., T.Z. Strybel, and R.W. Proctor. 2006. Effects of displacement magnitude and direction of auditory cues on auditory spatial facilitation of visual search. *Human Factors: Journal of the Human Factors and Ergonomics Society* 48 (3): 587–599.

Wang, L., and M.A. Hart. 2005. Influence of auditory modeling on learning a swimming skill. *Perceptual and Motor Skills* 100 (Part 1): 640–648.

Wellner, M., A. Schaufelberger, J.V. Zitzewitz, and R. Riener. 2008. Evaluation of visual and auditory feedback in virtual obstacle walking. *Presence: Teleoperators and Virtual Environments* 17 (5): 512–524.

Wenzel, E.M., M. Arruda, D.J. Kistler, and F.L. Wightman. 1993. Localization using non-individualized head-related transfer functions. *J. Acoust. Soc. Am.* 94 (1): 111–123.

Wright, B.A., and Y. Zhang. 2006. A review of learning with normal and altered sound-localization cues in human adults. *International Journal of Audiology* 45 (7): 92–98.

Zahorik, P., P. Bangayan, V. Sundareswaran, K. Wang, and C. Tam. 2006. Perceptual recalibration in human sound localization: Learning to remediate front-back reversals. *J. Acoust. Soc. Am.* 120 (1): 343–359.

Zhang, Y., T. Fernando, H. Xiao, and A.R.L. Travis. 2006. Evaluation of auditory and visual feedback on task performance in virtual assembly environment. *Presence: Teleoperators and Virtual Environments* 15 (6): 613–626.

Zychaluk, K., and D.H. Foster. 2009. Model-free estimation of the psychometric function. *Attention, Perception and Psychophysics* 71 (6): 1414–1425.

Seki, Y., and T. Sato. 2011. A training system of orientation and mobility for blind people using acoustic virtual reality. *IEEE Transactions on Neural Systems and Rehabilitation Engineering* 19 (1): 95–104.

Shams, L., and A.R. Seitz. 2008. Benefits of multisensory learning. *Trends in Cognitive Sciences* 12 (11): 411–417.

Shelton, A.L., and T. Smith. 2002. The influence of vision on the absolute identification of sound-source position. *Perception and Psychophysics* 28 (4): 589–599.

Sigrist, R., G. Rauter, R. Riener & P. Wolf. 2013. Augmented visual, auditory, haptic, and multimodal feedback in motor learning: a review. *Psychonomic Bulletin and Review* 20 (1): 21–53.

van der Heijden, K., and E. Formisano. 2013. Using noise to compute sound-source positions: model by. *Hearing Research* 30 (3): 35–40.

Wenzel, E.M. 1992. Localization in virtual acoustic displays. *Presence: Teleoperators and Virtual Environments* 1 (1): 80–107.

Section III

Digital Representation
and Modeling of Skill

9 Digital Representation and Modeling of Human Skill

Carlo Alberto Avizzano

CONTENTS

INTRODUCTION

This chapter presents novel digital representation techniques that take into account the ability to capture the human sensitivity to environmental variation and its ability to adapt motion behavior to compensate and regulate external processes. The chapter is inspired by early studies of Yokokohji (1996, "What you see is what you feel") who argued how multimodal environments could be used to simulate the dynamic of physical processes and intervene on learning. At the basis for motion learning research, we assume that an internal (at least simplified) model for motion exists (Flash, 1985; Kelso, 1984), but the approach also benefits the efforts of Henmi (1998) and Sano (1999) who conceptualized how motion models could be extracted and trained from the observation of user skills.

This problem is shared in robotics, those known as programming by demonstration (Billard, 2008), imitation learning (Azad et al., 2007), or motion programming (Wren and Pentland, 1998). The related models presented in the literature typically use rewarding approaches (Kober and Peters, 2009; Kormushev et al., 2010; Pastor et al., 2009) or probabilistic instruments (Wei et al., 2011; Wang and Fleet, 2008) that converge to a solution that maximizes the motion expectations. Before designing our approach we examined relevant state-of-the-art solutions (Avizzano, 2011) and found some limitations in the representation of skills for almost all of them:

- SLDS: Switching Linear Dynamic Models (Pavlovic et al., 2001)
- GMR: Gaussian Mixture Regression (Calinon et al., 2006)
- DMP: Dynamic Motion Primitives (Ijspeert, 2001)

- Local Model Variants (Atkenson, 1990)
- Gaussian Processes Variants (Wang and Fleet, 2008; Ye and Liu, 2010)
- SEDS, Stable Estimator Dynamical Systems (Khansari-Zadeh and Billard, 2010)
- Motion-Predictive Control (Da Silva et al., 2008)

Our model solves most of the criticality found for these models but could also be operated as a benchmark methodology to assess the level of experience achieved by the user in a way that is analogous to what has been presented by Li (2008) for the recognition of signatures in human motions. The proposed approach heavily relies on the ability to decompose complex interactive motion into regular elements that can be segmented and labeled through the user of a standardized interaction framework (Avizzano et al., 2009), and the ability to intervene in the reverse composition and synthesis of these motions through a multilayered/supervisory controller that could be related to human "will," its "consciousness," and its physical abilities (Warren, 2006).

A focus on multistyle ball "launch" and "catch" exercises is taken to explain mathematics and properties of the proposed approach. For this purpose a simplified experimental platform has been developed and employed to collect the style of motion properties (Figure 9.1).

MOTIVATION FOR A NEW TOOL

The existing state-of-the-art tools have been analyzed in order to understand the potentialities to properly model skills behaviors. In particular it has been observed that all the existing systems are arranged to learn from samples, but miss to capture

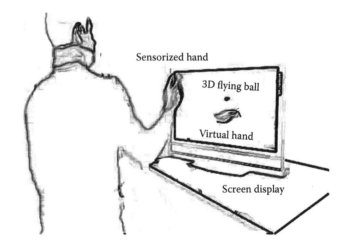

Sensorized hand

3D flying ball

Virtual hand

Screen display

FIGURE 9.1 The setup of the motion experiments. The hand position is captured via Polhemus sensor, and the ball trajectory simulated with real-time physics. The experiment of throwing and catching is repeated several times while eight state variables are monitored (hand/ball X/Y positions and velocities) at a frequency of 250 Hz.

the relationships between the whole motion shapes and the relevant environmental conditions that influence such changes. We addressed this weakness as *lack of adaptation* during real task generation. This miss is critical when the generated motions are related to other external entities such as objects or underlying control processes that could produce some changes in the motion profiles.

In addition, most models are reliable only when checked within the regions of space that have been used during the training process (Vijayakumar, 2005; Khansari-Zadeh and Billard, 2010). It is not uncommon that marginal stability on unstable behaviors emerges in other regions of the space. To improve the adherence between a parametric model and collected data, Grollman and Billard (2011) introduced some penalty-style algorithms for taking advantage also from wrong demonstrations. The *unstable behavior* of these systems is partially solved with some intrinsically stable models (such as in the DMP approach) or training procedures (such as in the SEDS approach).

Another difficulty in the use of the existing algorithm is the "abuse" of the derivative operation to generate a source for learning and regression. Such use is acceptable when theoretical models should be derived from well-shaped data, but this typically generates a lot of *differentiation noise* with real data.

Some models also *lack introspection* capabilities, in the sense that the model they generate may only be used as a black-box (Wang, 2008). This condition is particularly critical when, such as in the case of skills generation and control, one requires the model to be manipulated in order to meet specific learning requirements, such as manipulation of physical conditions, manipulation of geometrical constraints, and addition of artificial effects and noise.

Most models do not offer calibration procedures to adapt the internal complexity of the model to the one required for the skill under consideration. As a consequence, typical *oversizing* and *overfitting issues* appear in models. In Figure 9.2 we reworked an example from Calinon et al. (2010) to ask if DMP overfits a simple trajectory, with more primitives (32 instead of 8).

FIGURE 9.2 corrDMP example (Calinon, 2010), inflated with 32 primitives instead of 8. The attractors diverge from the reference trajectory and only their combined behavior allows motion to stay in the exemplary motion.

DMP attractors, instead of being more and more adherent to the reference geometry, diverge from that, thus achieving convergence on the desired trajectory only by reciprocal cancellation of dynamic effects.

Another critical feature of existing algorithms is the absence of relationships among the energy of the performed trajectories and the energy embedded in the model state. Hence there is high *energy consumption* used during the motion generation which is not respective of the energy used by muscles in controlling the same motion.

Finally it was found that all the proposed algorithms miss exploiting a final model in such a way that it is ready for further, or simultaneous, optimization and constraint exploitation. This feature that can be considered only marginal in batch and graphical simulation is of critical importance in the case of skills modeling systems. In some specific learning approaches, reinforcement learning is adopted in batches to meet specific constraints/optimization issues (Kormushev et al., 2010), but the solution found by this algorithm often produces trajectories that are far from typical motion coordination embedded in the examples.

COMPUTATIONAL PIPELINE AND DATA MANAGEMENT TOOLS

To handle the complexity of skills data, a rich development framework that facilitates the data manipulation has been developed: the framework makes use of a hybrid integration between Matlab®, Simulink®, Python™, SQLite, and C-Code to achieve an optimal analysis setup. The framework also ensures portability between most common operating systems (Linux/Unix, MacOS, and Windows).

The framework support has several features:

- It facilitates the collection, labeling, and reuse of experimental data through the assistance of DBase architecture based on SQLite (Owens, 2006). The interface provided to the user allows the user to merge and manage experiments coming from several experiments or data sources. This component within the system architecture is called "DataKit."
- Second, wherever useful, the system facilitates integration with the multimodal virtual environment (VE), by allowing support for real-time data collection, data storage and analysis, and the processing of a set of benchmarks that can be employed for training sessions. Question-related data organization, performances, VE integration, simulation, and real-time training have been discussed elsewhere (e.g., Avizzano et al., 2010).
- Third, the framework offers a set of computational kits that provide different levels of analysis and representations. In particular the computational framework provides the following features: source control, filtering, task decomposition, (de)featuring, learning, and simulating.

In Figure 9.3, the software architecture is summarized. This architecture is required to interact, analyze, and train physical skills. Most operations are managed to track automatically the data hierarchy in the generation by monitoring and recording data source, generation algorithm, and parameters used for capture. Figure 9.3 highlights most of the phases that have been supported within the framework such as data

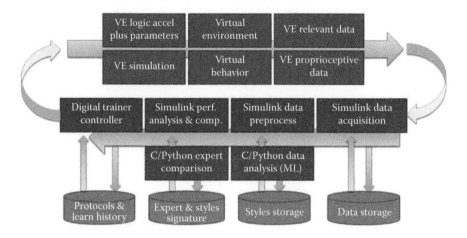

FIGURE 9.3 Computational and analysis framework. (From Avizzano, C.A., Ruffaldi, E., and Bergamasco, M. 2010. Digital management and representation of perceptual motor skills. In *Proceedings of Third Conference on Human Factors and Ergonomics,* pp. 863–872. Boca Raton, FL: CRC Press/Taylor & Francis.)

reading and acquisition, feature extraction, filtering, segmentation, labeling, warping, latent projection, and learning. Most of the components could be interchanged to set up the optimal computation pipeline; for instance, the user could apply several possible filtering strategies, each with pros and cons (Lippi et al., 2011).

One of the most important phases for correct learning is a proper segmentation and labeling. Labeling interprets motion according to specific data properties and associates a name to each motion. The temporal sequences here identified are named *chunks.* Chunks labeled with the same name share homogeneous information and represent the base of data for the model generation.

The learning process focuses on separate phases found by the labeling process, demanding to higher supervisory controllers the role of reassembling phases to achieve structured motions. To each specific phase a set of hyperparameters will be associated in a vector **H**. The hyperparameter vector collects in a unique set those variables that are considered to be strictly related to the profile of motions. It is important that the model learning process identifies the variability of the generated motions with respect to the hyperparameters. Typical hyperparameters can be the boundary conditions (position, velocities, or combination of them), time/frequency of execution, or level of forces. In the examples that follow, the differential of boundary positions will be used as a hyperparameter vector.

DEVELOPING A DIGITAL REPRESENTATION FOR A SKILL MODEL

With these objectives and scenario in mind, a representation model that employs the presented framework and overcomes some major issues of existing models has been developed. The model assumes that the motion trajectories generated by the user have been organized into sets of (generalized) coordinates or trajectories that are representative for the task modeling:

$$\mathcal{M} \rightarrow \begin{cases} q_{1,1}(t), q_{1,2}(t), q_{1,3}(t), \dots \\ q_{2,1}(t), q_{2,2}(t), q_{2,3}(t) \\ \quad \dots \end{cases}$$

Here the first subscript is associated with the coordinate, and the second (only indicated when relevant) to the repetition example achieved using segmentation and grouping of structured motions (Avizzano et al., 2009). Without loss of accuracy the motion analysis can be carried out using "function expansion series" which are already used in mathematics and other scientific fields for arbitrary function approximation. In particular a function set that employs a combination of sinusoidal and linear components has been employed. The linear behavior has been chosen to facilitate the matching of the boundary conditions, while the harmonic components have been chosen for energy issues and computational efficient tools (discrete sine transformation, DST). The function set preserves relevant features such as orthogonal properties with respect to the product integral, and decomposition between boundary values (guaranteed by first two components and other derivatives constraints). The general form of the model for the k generalized coordinate could be expressed as

$$\hat{q}_k(H, \tau) = \sum_{i=1}^{n} p_i(H)\Phi_i(\tau) = P^T(H)\Phi(\tau)$$

where the hat here stands to indicate that there is an approximation due to the truncation of the series, H is the hyperparameter vector, P is the expansion coefficient vector, τ is an experiment normalized time ($0 < \tau < 1$), and Φ is the *function set* vector defined as one of the below sets:

$$\Phi_1(t) = \left[1, \tau, \cos(2\pi\tau), \sin(2\pi\tau), \cos(4\pi\tau), \sin(4\pi\tau), \dots, \cos(n2\pi\tau)\right]^T$$

$$\Phi_2(t) = \left[1, \tau, \sin(\pi\tau), \sin(2\pi\tau), \dots, \sin(n\pi\tau)\right]^T$$

that is a DST enriched with a linear behavior that helps us to introduce two relevant characteristics in the computational pipeline:

1. Reduction of discontinuities on points $\{-1, 0, 1\}$. Reduction of discontinuities helps to cancel the Gibbs effect (Davis, 1963) that typically makes inadequate the use of pure harmonic decomposition
2. A computational efficient numerical algorithm based on the boundary values for the linear approximation and on FFT manipulation for the harmonic components (Figure 9.4)[*]

---

[*] The reason for having introduced a second set Φ_2 is to ensure the control of differential conditions at boundaries. The periodicity of Φ_1 imposes that $\nabla\phi_1(0) = \nabla\phi_1(1)$ whatever would be the value of the coefficients. The choice of ϕ_2 removes this limitation but also requires more accurate data handling to achieve proper expansion. In what follows we proceed in the analysis using Φ_1, while later, when specified, Φ_2 will be adopted.

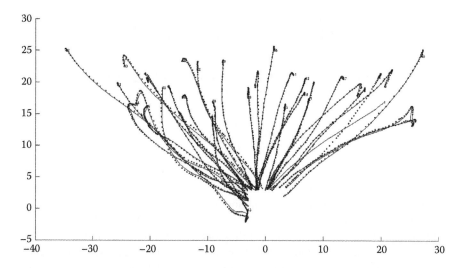

FIGURE 9.4 Function approximation (n=5), each motion trajectory here has been approximated with a linear trend, plus three harmonic components which store the particular motion profile.

The representation through function expansion has some similarities to the GMR (SEDS) or DMP approaches in the sense that it models the dynamic evolution of a system with the summation of several simple components that are retrieved from the analysis of motion data. However, in our case the set of functions used to generate the motion behavior do not activate sequentially with time (as for DMP) or spatially (GMR approximation), but always contribute to the complete motion simultaneously. These two approximations have been employed in DMP and SEDS, in order to reduce the interference among the simple models and facilitate their computation from exemplary data. In the function expansion models a similar feature is achieved by insertion of the orthogonal property of the functions. In addition the use of harmonic-related components allows us to have a direct estimation of the energy associated to each approximation. In particular when $n \to \infty$ the error energy goes to zero, and the energy of the harmonic components can be estimated as

$$\sum_{i=3}^{\infty} p_i^2;$$

hence, the energy of the error introduced by truncating the series to n can be computed as

$$\sum_{i=k}^{\infty} p_i^2$$

or (by difference) as a function of the energy in the series:

$$\varepsilon_{k,i} = \int \left(q_{k,i}(\tau) - \hat{q}_{k,i}(H,t)\right)^2 dt = E(q_{k,i}) - \boldsymbol{P}^T\boldsymbol{P}$$

where E is a property of the motion trajectory not depending on the motion model and computed as the energy of the motion by its linear trend:

$$E(q_{k,i}) = q_{k,i}^2(0) + \left(q_{k,i}(1) - q_{k,i}(0)\right)^2 + \int_0^1 \left(q_{k,i}(\tau) - (q_{k,i}(0)(1-\tau) + \tau(q_{k,i}(1)))\right)^2 d\tau$$

The ability to use the vector norm to estimate the function approximation is fundamental in the design of the approximation function and in the elaboration of adaptation algorithms that are based on the optimization of the likelihood.

In the following text the case of juggling one ball will be considered. A special exercise performed by launching one ball in the air and catching it back (setup shown in Figure 9.1) will produce data for motion analysis. A Polhemus sensor is attached to the hand back to record XYZ (lateral, vertical, frontal directions) positions of the hand in space. A simple physically based model of a ball dynamics was implemented in the virtual environment. Motion segmentation and labeling were implemented by checking relative displacement between the hand and the digital ball (the same operation can be performed using more specific motion analysis algorithms). Two generalized coordinates were used in order to directly map to the Cartesian position of the hand ($q_1 \rightarrow x_{hand}$, $q_2 \rightarrow y_{hand}$).

In order to generate appropriate information for the model design using experimental data, data were collected from sensor observations and divided from long exercises into short time sequences that were associated to specific motion phases. Then short motion phases were grouped together and transformed through the function-set series expansion described above. In such a way, a mapping between motion trajectories and points in the function space was achieved:

$$q_{1,1}(t) \rightarrow \boldsymbol{P}_{1,1}$$
$$q_{1,2}(t) \rightarrow \boldsymbol{P}_{1,2}$$
$$\dots$$
$$q_{1,r}(t) \rightarrow \boldsymbol{P}_{1,r}$$

In Figure 9.4 motion points have been sampled during the return phase and approximated in the continuous line using a fifth-order function expansion. For simplicity in the segmentation, return trajectories were cut when the user entered a squared area close to the resting position (chosen as origin). In this experiment, two hyperparameters have been considered relevant to characterize the motions δx and δy, defined as follows:

$$H = \begin{pmatrix} \delta x \\ \delta y \end{pmatrix} = \begin{pmatrix} x(1) - x(0) \\ x(1) - y(1) \end{pmatrix}$$

where x and $y \rightarrow (q_1, q_2)$, represent the horizontal/vertical motions with a normalized motion time in such a way that each trajectory starts in 0 and terminates in 1. The variation of the **P** vector with respect to the hyperparameters was analyzed by plotting each component of **P** against δx and δy.

With surprise, it was found that most components have a very close linear relationship against the hyperparameters, which provided us an idea to design a model that generalizes the motion trajectories. Hence, in order to proceed, an initial assumption was made that the **P** vectors were generated by a memory-less Gaussian process that depends only linearly from hyperparameters. This hypothesis implies that residuals from the fittings should respect a normal distribution. Figure 9.5 shows a plot of the typical scenario that is achieved through a plain linear regression (no partial least squares [PLS] or robust fitting). The regression planes have been achieved using the **Φ₁** dataset. Each plane represents how the **pᵢ** components vary when different boundary conditions (that are related to the hyperparameters) are requested.

The residuals between the points and the planes can be associated to several factors:

1. Noise in the captured data
2. Natural intervariance of human motion when identical boundary conditions are requested
3. Outliers in trajectories (bad gestures, bad segmentations)
4. Not relevant to the specific components with respect to the specific motion (usually the associated component values are small)

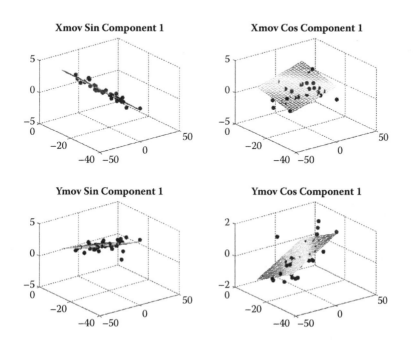

FIGURE 9.5 Plot of P_i values against the hyperparameters and fitting with a linear regression.

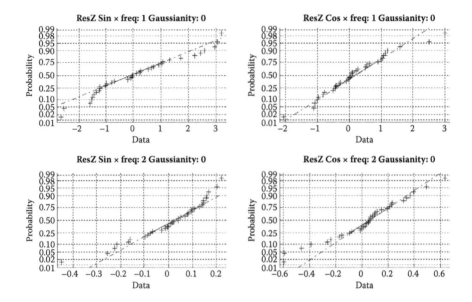

FIGURE 9.6 Analysis of residuals distribution, of X motions, first and second harmonic. The distribution with respect a Gaussian trend is considered: Lilliefor test (in the headings) confirms the Gaussian hypothesis.

A statistics of these residuals has been examined with a Lilliefor test to verify the Gaussian process hypothesis underlying the modeling process (Figure 9.6). In all cases where a strict correlation between motion and hyperparameters was found, the normality test was successful.

The presence of highly correlated fitting planes for the approximating decomposition, as well as the presence of constant and uncorrelated factors, suggests that employing a reduced order representation may help to achieve a more performing and compact representation. There exist a wide variety of tools to perform this passage in a more accurate and coherent manner (i.e., taking combined effects into consideration), both from the probabilistic framework (Rasmussen and Williams, 2006) and from the statistic framework (one for all the local linear embedding) (Roweis and Laurence, 2000).

In order to understand what happens to motion when the **P** vector is projected on the regression hyperplane, we represented a comparison of trajectories in Figure 9.7. The figure shows on the left the real motion points (in bullets) and the trajectories achieved when using motion interpolation with the full expansion series (left side). The numbers on the trajectories represent the sequential order of the motion captured during the experiment. On the right, the ad hoc interpolation was replaced with the respective points extracted from the regression planes and constrained to the hyperparameters determined from boundary conditions. This analysis was performed on the data as they are, without any outlier removal (prior regression or plotting) and without having a precise segmentation of trajectories.

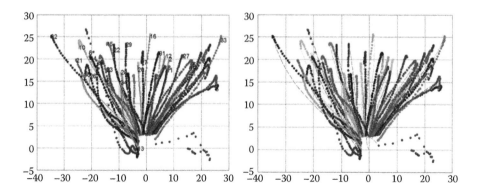

FIGURE 9.7 Comparison between the interpolation provided per gesture, and the equivalent model regressed from data using the coefficient linearity observed in Figure 9.5. Trajectories are now less adherent to motion, but style of the gesture is still preserved.

The motion generation performs satisfactorily, the profiles of motion are maintained, and the overall two-dimensional (2D) geometry respects those generated by human gestures. The major discrepancy is found at trajectory start and is generated by having "violated" the initial condition on the single trajectories using the planes instead of the specific trajectory points. The model for the skills motion, even without optimizations, is very compact: $2 \cdot (n-2) \cdot (h+1)$ parameters where n is the order of the expansion, and h is the dimension of the hyperparameter space.

Figure 9.8 shows a full model fifth-order model having $2 \times 3 \times 3$ parameters to encode motion and tabulates the confidence intervals (at 95%) estimated for parameters. The projected model simplifies and regularizes the motion while producing a repeatable and predictable spatial property as a sole function of task hyperparameters.

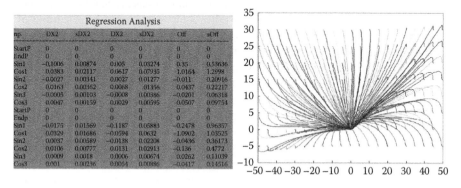

Regression Analysis					
np	DX2	sDX2	DX2	sDX2	Off	sOff
StartP	0	0	0	0	0	0
EndP	0	0	0	0	0	0
Sin1	−0.1006	0.00874	0.005	0.03274	0.35	0.53636
Cos1	0.0383	0.02117	0.0617	0.07935	1.0164	1.2998
Sin2	−0.0027	0.00341	0.0027	0.01277	−0.011	0.20916
Cos2	0.0163	0.00362	0.0068	.01356	0.0437	0.22217
Sin3	−0.0005	0.00103	−0.0008	0.00386	−0.0201	0.06318
Cos3	0.0047	0.00159	0.0029	0.00595	−0.0507	0.09754
StartP	0	0	0	0	0	0
Endp	0	0	0	0	0	0
Sin1	−0.0175	0.01569	−0.1187	0.05883	−0.2478	0.96357
Cos1	0.0329	0.01686	−0.0594	0.0632	−1.0902	1.03525
Sin2	0.0037	0.00589	−0.0138	0.02208	−0.0436	0.36173
Cos2	0.0106	0.00777	0.0131	0.02913	−0.136	0.4772
Sin3	0.0009	0.0018	0.0006	0.00674	0.0262	0.11039
Cos3	0.001	0.00236	0.0054	0.00886	−0.0417	0.14516

FIGURE 9.8 Regressed parameters from a fifth-order model, and its relative spatial properties. On the left we represented the parameter values in function of the expected motion (delta move X,Y) and the errors on the parameters to include that achieve ($p < 0.05$). On the right we plot the motion generated from the acquired samples in a workspace that is 4 times bigger than the observed samples. The system shows good properties of maintaining simultaneously stability and generalization.

MODEL DISCUSSION

The proposed model based on the function expansion series is an intrinsically stable model, because it models hyperparameter adaptation through a linear property. This property allows the designer to generalize motion over a large portion of space (even beyond the observed region) without introducing stability issues. Notwithstanding this simple underlying linear behavior, the model has shown to be flexible to automatically capture several features of the generalization such as translations, stretching, rotation, axial, and central symmetries.

The model abstraction has been possible by warping all the data collected in the experiments into a unique *virtual time* interval [0,1] in which each phase is assumed to be executed. This compression is common to several other approaches and does not introduce a relevant limitation in the expressiveness of the model because it can be removed with auxiliary state variables that model the virtual time (Pastor et al., 2011).

In contrast to motion studies that collapse or reduce intermotion variability to pure stretching operation by appropriate scaling and translation of Cartesian trajectories, here the variation taken into consideration is modeled with respect to any set of hyperparameters and produces effects on a chosen relevant set of generalized coordinates. The components of the motion vector **P** can be determined using the offset $\mathbf{P_0}$ and gradient $\nabla \mathbf{P}$ estimated during the regression:

$$P_i(H) = P_{0,i} + \nabla P_i \cdot H$$

$$q_i(\tau) = P_i^T(H)\Phi(\tau)$$

The above formula provides a means to reconstruct motions with any two boundary constraints (**H**). In particular, while boundary conditions are automatically and exactly satisfied by the linear behavior, the other components capture the features of the motion profile. In Figure 9.9, four different styles of motion were recorded, these motions were produced by simply asking the performers to exhibit particular shapes in the air while catching a floating ball, the same hardware of Figure 9.1 and modeling algorithms described before were employed to collect and manipulate data. Again, the virtual ball to be captured was thrown in the air by the software with random initial parameters in order to solicit the variability of the user response. The produced styles were named for clarity as *umbrella, straight, curled,* and *circular* and showed in the figure how all the essential features of these different motions can be captured and generalized by the proposed model. The graphs report on the left side what has been recorded by the digital system, and on the right side, the responses that could be generated from the learned models when triggered by hyperparameters changing

FIGURE 9.9 *(See facing page)* Exemplary motion styles modeled with same regression algorithm. Sample experimental data are represented on the left, achieved models on the right. XY axes are positions expressed in centimeters. Trajectories have been modified in order to start (catch) or terminate (return) at point (0.0). From top to bottom: counterclockwise catch, counterclockwise return, curled catch, straight catch, straight return.

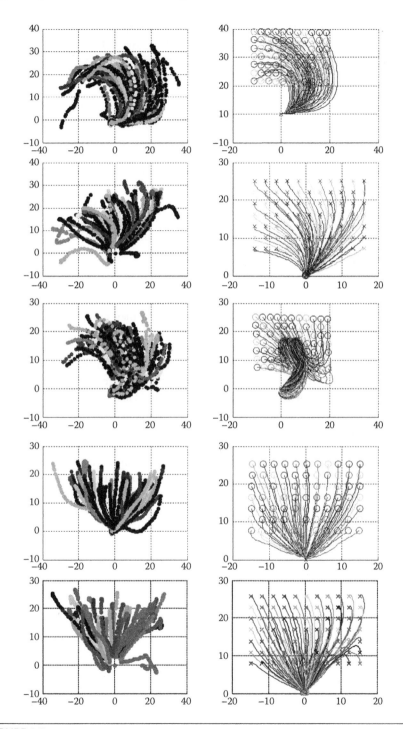

FIGURE 9.9

on a grid. Learned motions show several interesting properties such as robustness to segmentation alignments, robustness to motion outliers (false trajectories), automatic identification of symmetries, stretches and rotations, implicit regularity of motion caused by indirect averages with regression, intrinsic stability by construction, and whole space solvability.

In the proposed model, *over-fitting issues* are naturally limited by the implicit energy limitation of the design procedure and can be monitored and controlled by scoring the approximation performances.

The model generation always makes use of the complete trajectories to determine the shape of the model trajectories. This feature ensures a very strong tolerance to noise. In such a way, and to an advantage with respect to other model generation methods, our model makes added value from bigger data collection while it remains efficient and stable even when small datasets are available. In addition the model does not require any data differentiation (which enhances the noise feature) to produce a proper fitting.

ADAPTATION

Modeling is not per se sufficient to express the complexity of a skill. Humans have the intrinsic ability of being able to adapt the motion they have learned to different circumstances, and to control how to replan such a motion when some alteration from the previous plans emerges from environment observation. This ability to modify in real time the original motion plans allows humans to adapt to context changes and be more tolerant and robust when interacting in real conditions. In what follows we will show how to implement this capability in an optimal but still numerically efficient way.

In our models we will consider only two types of plan changes: replan due to control errors and replan due to target changes. In the first case it is assumed that notwithstanding what has been planned, at a certain point, during the execution of a motion, the system (the human) detects that his position differs from the expected one, and therefore he needs to redesign his motion program in order to match the original objective; in the second case, during the execution of the motion program, the human (or the system) detects that the final objective has changed in the meanwhile, and therefore he needs to find a new motion program that is similar to the previous, continuous in the point of observation, and such that it matches the final target.

Now suppose that at a given time τ_1 $(0 < \tau_1 < 1)$ a replan of the motion trajectory is needed, in the generalized coordinate i (we omit the subscript i for clarity here)

$$q^*(t) = P^{*T}\Phi(\tau) \qquad \tau \in [\tau_1, 1]$$

Subjected to the following constraints:

$$C1:\ q^*(\tau_1) = q(\tau_1) + \Delta q_{\tau_1}$$

$$C2:\ q^*(1) = q(1) + \Delta q_f$$

By substituting T1 in C1, C2, and rewriting the problem in the $\delta P = P^* - P$ parameter vector, we obtain with simple algebra manipulation that the two constraint conditions reduce to

$$\delta P_i^T \Phi(t_1) = \Delta q_{\tau_1}$$

$$\delta P_i^T \Phi(1) = \Delta q_f$$

Hence the problem of finding a new P^* that satisfies the constraints while maintaining the maximum likelihood with the given trajectory can be achieved by minimizing $|\delta P|^2$ under the above constraints through the use of Monroe-Penrose Pseudoinverse:

$$\delta P_i^T = \begin{bmatrix} \Delta q_f & \Delta q_{\tau_1} \end{bmatrix} \begin{bmatrix} \Phi(1) & \Phi(\tau_1) \end{bmatrix}^\dagger$$

Thanks to Parseval's equality, the minimization of the trajectory error here is performed through the minimization of the harmonic coefficients. When applied to our expansion set and thanks to the choice of orthogonal functions, the maximum likelihood problem translates in a minimum problem on $(\delta P^T \, \delta P)$. Even if the use of the pseudoinverse with a minimum distance norm does not preserve the hyperplane constraints determined through the linear regression on $\mathbf{P}$ coefficients, if the corrective errors are expressed in a small neighborhood, the distortion induced on the trajectory is not significant with respect to the approximation associated to the regression itself. In other cases a third constraint equation should be introduced:

$$\text{C3: } \delta \overline{P} = \nabla P \, \delta h$$

where δh represents the reduced-order motion vector in the "free" hyperparameter space. With this substitution, the solution assumes the form of a generalized-constrained pseudoinverse:

$$\delta \overline{P}^T = \delta h^T \, \nabla P^T = \begin{bmatrix} \Delta x_f & \Delta x_{t_1} \end{bmatrix} \begin{bmatrix} \nabla P^T \Phi(1) & \nabla P^T \Phi(t_1) \end{bmatrix}^\dagger \nabla P^T$$

As we will see below this condition has less importance when a small deviation from the original trajectory is considered, but it assumes a real value when constraints on the derivative condition are assumed.

In Figure 9.10 the effects of the adaptation procedure are shown: an error detected right in the middle of a trajectory (left example) or to the target position (right). The arrows represent the error and the time on the trajectories, and the dotted trajectory is the adapted motion. The adaptation recovers the motion in a smooth way, maintains the overall trajectory profile, and does not introduce sharp variations on the running motion.

A major limitation of the Φ_1 function set is the control of differential properties at boundaries. This set of functions has derivatives that exactly repeat with period 1; hence, there is no possibility to fix differential properties in 0 different from those in 1.

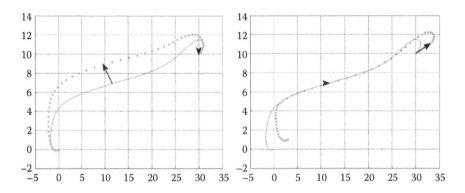

FIGURE 9.10 Various adaptation profiles for errors on the tracking point or on the final objective.

This property is of critical importance to specify the constraints for particular motion gestures that should match kinematic conditions at boundaries (for instance, ensure a given final condition, or a nonzero final condition with a zero starting motion). A particular case for splining trajectories is the ability to introduce additional constraints on the final point velocity, such as

$$C4: \dot{q}(1) = \dot{q}_f = P^T \Phi'(1)$$

In order to circumvent such an issue we introduced a Φ_2 whose period is 2 ($[-1,1]$). This function set allows different differential conditions at points 0 and 1 but needs to be trained with symmetrical datasets as shown in Figure 9.11. The training shape extends the reference shape with a symmetric profile with respect to the point {0,0}. The figure represents with the dashed line how the trajectory extension is achieved. Finally, being this function "Odd," it allows the harmonic decomposition to be based only on an odd functions set (sine functions) and does not augment the number of parameters required to achieve the same level of accuracy.

In adding a differential condition, the combined use of C3, C4 is critical in order to avoid distortion of the geometrical shapes (Figure 9.12). The data in the figure have been generated from a segmentation algorithm that divides chunks when the vertical motion of the hand changes its direction (null Y velocity). A simple extrapolation/generalization of the data trajectory loses this property (top, right), while the introduction of the velocity constraint determined uniquely with a minimum energy principle (P norm) creates several ripples and oscillating effects (bottom left). Only the combined effect of the C3, C4 allows proper generalization of the motion (bottom right).

ADAPTATION DISCUSSION

One of the relevant features of the model based on expansion function series is that it maintain a mathematical form that can be fitted to data without the requirement of differentiation, but it offers analytical differentiation abilities in the final form. This property has been used to adapt to differential boundary conditions in the previous section. In some tools, the *lack of adaptation* strategies during real task generation

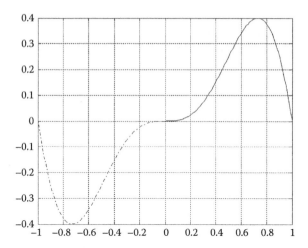

FIGURE 9.11 Symmetrization of the chunk is used to allow different boundary condition on learned shapes while preserving the model complexity.

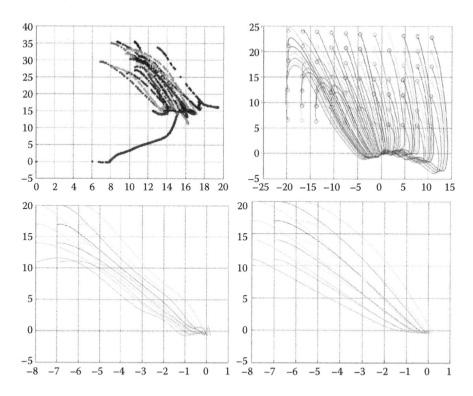

FIGURE 9.12 Trajectory generation with constraints: original data (top left), simple generalization (top right), minimum norm velocity constraint (bottom left), minimum norm and regression constraint (bottom right).

was compensated by the introduction of batch techniques that reshape the motion while losing the likelihood properties. Our adaptation algorithms provide an alternative real-time way of shaping the motion, and a powerful instrument to supervise at higher control levels the motion commands. The smooth properties of the alternate trajectories allow a direct forwarding of this model as a reference to a controller. Two types of adaptation tools were provided: a current position compensation that is useful in those cases where the correct position tracking cannot be ensured in the control and its value is progressively estimated during the motion (such as in the case of Kalman filtering, sensor fusion, drifting, underactuated systems, etc.); and a target position update that is useful in the case that when a motion is started only a rough estimation of the final position is available, and this estimation is refined step by step. The choice of $\mathbf{\Phi}$ sets also ensures that the solution to the inversion problem is always consistent and does not degenerate if $\tau_1 \neq 1$.

Another relevant feature of the proposed methodology is its intrinsic stability. The proposed method is not only stable in a strict region around the training points, but it also maintains this property on the whole workspace. In contrast to existing techniques, the proposed model supports smooth deformations of trajectory geometries that automatically encode the spatial variability caught from the examples (mirroring point/lines, rotation, linear compression/expansions). This feature is otherwise introduced in existing models only with manual intervention of the designer. Finally, the regression methodology identifies motion planes that could define any specific trajectory matching each possible boundary condition.

Moreover, the presence of a clear and analytical function vector ($\mathbf{\Phi}$) allows exploration of the relationships between the generated signal and the design choices. One example is given by the simplicity of the adaptation procedure.

Adaptation and tracking issues are also essential properties in recombination of complex motions, where the complete behavior of a subject is achieved by assembling series of time limited trajectories that satisfy common boundary constraints on position and velocities. In real case applications, these boundary constraints are determined by the humans in the phases that immediately precede the end of the previous motion, and back influence the termination of this motion as well. The ability to set up a tracking adaptation algorithm that can simultaneously control position and velocity of the final point is crucial for proper assembly of these chunks.

We are currently running statistics to validate the existence of the motion hyperplanes on a larger motion set than the ones presented in this discussion, however, if confirmed, these hyperplanes can also be adopted as a powerful instrument to identify a precise motion signature for benchmark, comparison, guidance, and recognition applications.

REFERENCES

Argall, B.D., Chernova, S., and Veloso, M. 2009. A survey of robot learning from demonstration. *Robotics and Autonomous Systems* 57(5):469–483.

Atkenson, C.G. 1990. Using local models to control movement. *Advances in Neural Information Processing Systems* 2:316–323.

Avizzano, C.A., Ruffaldi, E., and Bergamasco, M. 2009. Digital representation of SKILLS for human robot interaction. *Robot and Human Interactive Communication Proc.*: 32–37.

Avizzano, C.A., Ruffaldi, E., and Bergamasco, M. 2010. Digital management and representation of perceptual motor skills. In *Proceedings of Third Conference on Human Factors and Ergonomics*, Kaber, D. and Boy, G. (eds.), (pp. 863–872). Boca Raton, FL: CRC Press/Taylor & Francis.

Avizzano, C.A. 2011. Human Motion Models for HRI: Part I. *Proc. of IEEE RO-MAN*, Atlanta, GA.

Azad, P., Asfour, T., and Dillmann, R. 2007. Toward an Unified Representation for Imitation of Human Motion on Humanoids, Robotics and Automation, International Conference on, pp. 2558–2563, IEEE, Rome, Italy.

Billard, A., Calinon, S., Dillmann, R., and Schaal, S. 2008. Robot programming by demonstration, in *Handbook of Robotics*. Chapter 59, Springer.

Calinon, S., Guenter, F., and Billard, A. 2006. On learning, representing and generalizing a task in a humanoid robot. *IEEE Transactions on Systems, Man and Cybernetics* 37:286–298.

Calinon, S. http://programming-by-demonstration.org/sourcecodes.php

Da Silva, M., Abe, Y., and Popovic, J. 2008. Simulation of human motion data using short-horizon model predictive control. In *Proc. Eurographics*.

Davis, H.F. 1963. *Fourier Series and Orthogonal Functions*. New York: Dover.

Esen, H., Yano, K., and Buss, M. 2008. Force skill training with hybrid trainer model. In *Proceedings of IEEE International Workshop on Robot and Human Interactive Communication*, 9–14.

Flash, Hogan. 1985. Coordination of arm movements: An experimentally confirmed mathematical model. *Neurology*

Grollman, D.H., and Billard, A. 2011. Do not as I do: Learning from failed demonstrations. *IEEE International Conference on Robotics and Automation*, pp. 3804–3809, Shanghai.

Henmi, K., and Yoshikawa, T. 1998. Virtual lesson and its application to virtual calligraphy system. In *Proceedings of IEEE International Conference on Robotics and Automation*, Leuven, Belgium. 2:1275–1280.

Ijspeert, A.J., Nakanishi, J., and Schaal, S. 2001. Trajectory formation for imitation with nonlinear dynamical systems, in *Intelligent Robots and Systems*. Proceedings. 2001 IEEE/RSJ International Conference on, volume 2}, pp. 752–757.

Kelso, J.A. 1984. Phase transition and critical behavior in human bimanual coordination. *Am J. Physiol Regulatory, Integrative and Comparative Physiology* 246:1000–1004.

Khansari-Zadeh, S.M., and Billard, A. 2010. Imitation learning of globally stable non-linear point-to-point robot motions using nonlinear programming. In *Proceedings of the IEEE/RJS International Conference on Intelligent Robots and Systems*, pp. 2676–2682, Taipei, Taiwan.

Kober, J., and Peters, J. 2009. Learning motor primitives for robotics. *IEEE International Conference on Robotics and Automation*, Kobe, Japan.

Kormushev, P., Calinon, S., and Caldwell, D.G. 2010. Robot motor skill coordination with EM-based reinforcement learning. *IEEE/RSJ International Conference on Intelligent Robots and Systems*, Taipei, Taiwan.

Lawrence, N. 2005. Probabilistic non-linear principal component analysis with Gaussian process latent variable models. *Journal on Machine Learning Research* 6:1783–1816.

Li, Y.F. 2008. On signature invariants for effective motion trajectory recognition. *International Journal of Robotics Research*, 27(8):895–917; doi: 10.1177/0278364908091678.

Lippi, V., Avizzano, C.A., and Ruffaldi, E. 2011. Filtering motion data through piecewise polynomial approximation, *BIO Web of Conferences*, volume 1, EDP Sciences.

Owens, M. 2006. *The definitive guide to SQLite*, Apress.

Pastor, P., Hoffmann, H., Asfour, T., and Schaal, S. 2009. Learning and generalization of motor skills by learning from demonstration. *Robotics and Automation*, 2009, *IEEE International Conference on*, pp. 763–768, Kobe, Japan.

Pastor, P., Kalakrishnan, M., Chitta, S., Theodoru, E., and Schaal, S. 2011. Skill learning and task outcome prediction for manipulation. *IEEE International Conference on Robotics and Automation*, pp. 3828–3834, Shangahi.

Pavlovic, V., Rehg, J.M., and MacCormick, J. 2001. Learning switching linear models of human motion. *Advances in Neural Information Processing Systems*, pp. 981–987.

Rasmussen, C.E., and Williams, C.K.I. 2006. *Gaussian Process for Machine Learning*. Cambridge, MA: MIT Press (available online at www.gaussianprocess.org).

Roweis, S.T., and Laurence, K.S. 2000. Nonlinear dimensional reduction by locally linear embedding. *Science* 290:2323–2326.

Sano, A., Fujimoto, H., and Matsushita, K. 1999. Machine mediated training based on coaching, in Systems, Man, and Cybernetics, 1999. IEEE SMC'99 Conference Proceedings. *1999 IEEE International Conference on*, volume 2, pages 1070–1075.

Vijayakumar, S., Souza, A., and Schaal, S. 2005. Incremental online learning in high dimensions. *Neural Computation* 17(12):2602–2634.

Wang, J.M., and Fleet, D.J. 2008. Gaussian process dynamical models for human motion. *IEEE Transactions on Pattern Analysis and Machine Intelligence*, 30(2).

Warren, W. 2006. The dynamics of perception and action. *Psychological Review* 113:358–389.

Wei, X., Min, J., and Chai, J. 2011. Physically valid statistical models for Human Motion Generation. *ACM Transactions on Graphics* (TOG) 30, 3, 19.2.

Wren, C.R., and Pentland, A.P. 1998. Dynamic models of human motion. In *Proceedings of Automatic Face and Gesture Recognition*.

Ye, Y., and Liu, K. 2010. Synthesis of responsive motion using a dynamic model. In *Proceedings of Eurographics*.

Yokokohji, Y., Hollis, R., Kanade, T., Henmi, K., and Yoshikawa, T. 1996. Toward machine mediated training of motor skills. Skill transfer from human to human via virtual environment. In *Proceedings of IEEE International Workshop on Robot and Human Communication* 5:32–37.

10 Data Management for Evaluation and Training in Virtual Environments

Emanuele Ruffaldi

CONTENTS

INTRODUCTION

The adoption of virtual reality (VR) systems for training and performance evaluation has a long history due to the ability of such systems to evaluate in real time trainee performance and to compute and provide feedback along a training protocol. The technological improvements and the reduction of costs of such systems have paved the way to more flexibility in system design, easier investigation in experimentation strategies, and increased functional realism with smaller effort. In parallel to hardware advancements, software frameworks have improved in the direction of more flexibility in supporting sensor and rendering systems reducing the typical complexity of VR system setup. There is an aspect that has received a less systematic investigation: data management for virtual environments (VEs). The focus of this chapter is on the peculiarities of data management for dealing with the evaluation of subject performance in skill training scenarios and the overall management of training. This application domain presents the unique convergence of general data management aspects with VE data and architectural elements and aspects of skill training.

The aim of this chapter is to present a system for dealing with data emerging from skill training scenarios in VE and the rationale behind it. The system is presented together with the application in the domain of a rowing training system based on VE. In particular, the chapter describes the following:

- An architecture for the workflow in training VE
- Data access and processing methodology
- Organization specific for skill management

The chapter is organized as follows: the section Objectives and Architecture deals with the context of the training in VE and the workflow around this type of system. This section also covers the architectural view of the proposed system. In the third section, the organization and manipulation of information are presented in the frame of skill management. The fourth section is dedicated to the case study of a rowing training system in VE.

OBJECTIVES AND ARCHITECTURE

The scenario envisioned in this chapter deals with the creation of training experiments in VE involving an iterative process of design, implementation, and testing that converges to a final version of the experiment. The experiment could then graduate to a long-term training protocol as part of a complete training product. The experimental phase is characterized by flexibility for the exploration of design decisions and by collaboration among designers. The overall process can be represented in terms of a research and development workflow specialized for training in VE. The phases of the workflow are the following, as depicted in Figure 10.1:

- *Design phase,* in which the training protocol or experiment is designed
- *Implementation phase,* in which the VE is being implemented and tuned by means of the previous design phase and with the support of capturing sessions in which expert data are acquired for performance evaluation modules
- *Execution,* in which the experiment or training is performed on a number of participants with real-time analysis of performance and recording of execution. Due to the fact that execution can be performed by different experimenters or coaches, an additional action is the *tracking* of execution
- *Analysis* is performed after *execution* for the evaluation of system and training effectiveness
- *Publishing* is a branch of the iteration in which results are being extracted from the analysis and presented

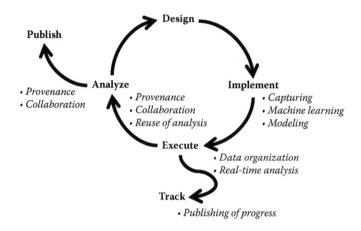

FIGURE 10.1   Workflow of the activities performed in data management.

Data management for experiments is recognized as an empowering tool for scientists (De Roure and Goble, 2009), and some general tools have been presented in the literature as Kepler (Jenkins et al., 2010), while others are focused on specific scientific domains, such as in the biological one (Saal et al., 2002). In the domain of the VE system, the focus of development has been on architectures for dealing with flexibility of the VR setup, graphical rendering, abstraction of sensors and models, and exchange of data among networked elements. The most successful frameworks are based on dataflow models like that of Figueroa et al. (2008) with the possibility of exchanging data among nodes in a flexible and efficient way. The X3D standard for virtual reality application is based on this model (Daly and Brutzman, 2007). This is in line with the publisher/subscriber approach common in robotics. Few works have dealt with the aspect of data sharing and evaluation like the one from Friedman et al. (2006).

A data management system covering the whole workflow can be described by two distinct elements. The first element deals with data representation and storage, allowing the exchange of information among the different phases of the workflow and among the components inside each phase. The second is the computational part that should be capable of supporting data processing in the different phases with the possibility of analyzing data in real time during the execution of the training and then offline.

ARCHITECTURE

The architecture of the system discussed in this chapter is aimed at supporting the activities of the above workflow as shown in Figure 10.2. This can be compared with other architectures for training like GVT (Gerbaud et al., 2008). The diagram presents the modules and the information exchanged among the modules in two separate modalities: real time and analysis.

The modality shown on the left in Figure 10.2 presents the real-time part that is active during the execution of evaluation and training sessions. In this modality

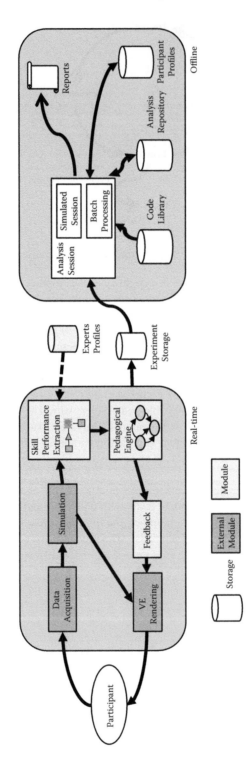

FIGURE 10.2 Architecture of the presented system with the left part expressing the real-time component, and the right part covering the batch analysis.

the focus is on the loop between the participant and the VE engine: the information acquired by sensors is combined with the simulation of the virtual world. The result is then used for extracting the performance indices and provided to the pedagogical engine that, based on the training protocol, controls the feedback. This architecture decouples also the VE rendering between two sources of stimuli: simulation and training feedback. The training feedback module behaves as an interface between the training engine and the specific VE framework, supporting also the reuse of the feedback strategies. This module and its integration with the pedagogical engine takes into account a previously discussed taxonomy of feedback in VE applied, for example, to sports training in VE (Ruffaldi et al., 2011).

PEDAGOGICAL ENGINE

The *pedagogical engine* is a key element of the architecture because it manages the training protocol by taking as input the performance indicators and controlling the overall system behavior. This module is a hierarchical state machine that deals with the structure of the training protocol at large, and represents each stage of the protocol as simpler event-driven state machine. The adoption of an event-driven state machine for representing simulations and protocols is recognized to ease the design and evaluation of the protocol, being in some way similar to the conceptual representation of the protocol (Sewell et al., 2004). Each state machine representing a block of the experiment is characterized by a set of input events and data that are directly produced by the VE or by means of performance extraction modules. The outcomes of the state machine are other events or variables that trigger feedback to the user. The aim of the overall system is to make it easier for field experts to map the protocol to a working system. Figure 10.3 shows the input and output elements with respect to the pedagogical engine.

PERFORMANCE EXTRACTION

The *performance extraction module* is a key feature of the discussed system because it has the objective to measure subject performance and to encode it in a compact and effective way. In particular this module performs a pipeline of activities over sensor data for filtering the signal and extracting the skill-related feature. These features can then be used together with machine learning components for performing recognition of patterns or model-based scoring. In the proposed framework these models are computed at design time based on previously captured data acquired on the training platform or from the real task. The extraction operation is also performed on data at different time scales depending on the domain, with it being common to analyze windows of data fixed over the last instants of time. In realistic cases both the time window and the feature extraction are performed based on a dynamic segmentation using pattern recognition.

OFF-LINE ANALYSIS

The right side of the architecture deals with the analysis part that is performed offline after the execution of the experiment or task. This part is employed both in

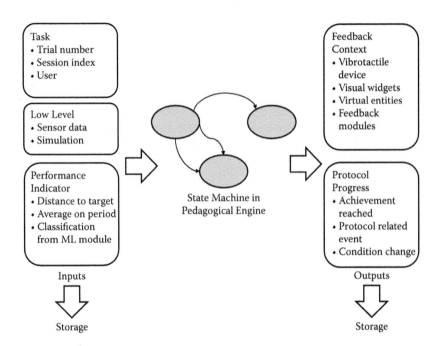

Inputs

Outputs

Storage

Storage

FIGURE 10.3 This diagram shows the information that is used to feed the protocol execution and the one that results as output. All these variables, considered as events, are stored in the task stream.

experimental analysis and in off-line duties in a production environment with complex scoring. The analysis is performed by accessing the data recorded during the training sessions and applying batch algorithms and simulated sessions. The aim is to provide a coherent view of what has been recorded and the possibility of extracting new models. In the design phase of the workflow, this part of the architecture is used for creating models from expert data, while in the postprocessing phase it is used for producing reports and publications.

INFORMATION FLOW

From the above architectural view, it emerges that most of the information is produced in a streaming way. Streaming information is generated during real-time execution or during simulation, and it is characterized by multiple streams of content as produced by sensors or by intermediate computing nodes. This type of information requires a data management component that takes into account the timing requirements of data production and preserves the context in which the information is produced. The system should manage the whole set of contextual information associated to experiments like the training protocol, training sessions, and users. As part of a comprehensive approach the system should be able to associate as contextual information also the specific model or software employed for performing the evaluation and the related parameters. Preserving this information is motivated by the need to recover the details of the models employed in the VE simulation for later analysis, in particular during the design iterations involving pilot experiments. This approach allows also performing playback of the acquired data in the VE. For this purpose a connection between experimental data and software model is needed, and employing a code versioning system is the natural choice. In particular in this work, a distributed versioning system like Mercurial or Git has been employed with the policy of performing a commit at the beginning of the training session and then storing the hash of the committed revision with the data being recorded.

INFORMATION STORAGE

Storage is an important aspect in VE for training because it provides a way to exchange information among the different modules and among different training sites. The objectives of storage are efficiency, data description, and integrity. Two main approaches can be identified in data storage: file oriented and database oriented. In the file-oriented approach content is manually organized by the system in a set of files and the management is delegated to the user. In database systems the application is provided with an interface for the access of data. The advantage of database systems is in their ability to provide structured and consistent ways for data backup, replication, and exchange over the network. Database Management Systems (DBMS) can be organized by the way they logically structure information. The most successful are the relational ones in which tables of uniformly organized information are related to each other with possible enforcement of the relationships. A different family of database systems is instead the one of object-oriented DBMS in which entities are described by objects belonging to a hierarchy of types and each object can refer to others. Manipulation of data in databases is provided by means of the Standard Query Language (SQL) that provides an almost uniform means for populating and querying information. This language is oriented toward traditional data management systems, and it has to be adapted and enhanced, together with the underlying database system, for supporting the type of information present in training VE environments. In particular these systems are not specifically tuned for large data matrices or large binary types. Although it is possible to adopt schemas in which the database stores reference external binary files, this approach, compared to the in-database storage of the binary content, does not allow the user to take advantage of the backup and replication capabilities of the database system. The other aspect that is partially limiting the adoption of classical database systems is the processing of information, in particular when data are flowing into the database from different sources. Research is moving in this direction by the adoption of streaming databases and processing engines (Arasu, 2004).

An interesting intermediate approach is the one provided by the Hierarchical Data Format 5 (HDF5) that is currently the most common standard for storing scientific data (Folk et al., 1999). HDF5 provides a logical representation of data made of a hierarchy of nodes associated with metadata. Leaves of the hierarchy contain data in the form of multidimensional numeric matrices while relationships among entities are supported by means of links. This format is quite efficient for the storage of information, and it can be interfaced with different languages and libraries. The logical structure of HDF5 is mapped by default to a single HDF5 file containing all the hierarchy, but it can be extended by means of a virtual file layer that is able to logically combine multiple files, map in memory data, and access networked data or virtual devices. This last feature is specifically welcome when the size of data is bigger than the maximum allowed by the file system. The main limitation of HDF5, if compared to databases, is in the domain of queries, specifically for expressiveness and performance, although some research has been performed in this direction (Gosink et al., 2006). The HDF5 poses few limitations on the type of content stored, and in some domains the logical structure and the metadata have been documented for sharing data (Dougherty et al., 2009).

In the discussed system the approach taken has been that of providing a higher-level Application Programming Interface (API) that is managing the storage of information by providing an abstraction over the storage mechanism. The adopted storage system is currently based on a series of SQLite databases. SQLite is an embedded database system that, different than other large relational databases, can be embedded inside a single application, and it provides enough flexibility for customizing storage and manipulation of data. A flexible schema allows representation of hierarchical information inside each SQLite database together with numerical data stored in large binary objects. The API allows hiding the structure of multiple databases in a way similar to the HDF5 virtual layer.

MANAGEMENT FOR SKILLS

This section deals with the management of skill-specific information both in terms of data representation and information processing.

Data Organization

From the above discussion there are different types of data elements involved in the workflow. The starting point are the data acquired from sensors and simulation that can be represented by a flow of events in which each typed event is marked by a time-stamp and content. When dealing with sensors, these events are typically uniform in type and uniformly spaced in time, for example, recordings from a motion capture system. Conversely, when a simulation produces data, these events are not synchronous and could contain different types of content. For example, in a medical VE involving bone drilling based on voxels, the number, positions, and type of voxels drilled are produced asynchronous. Variable events are also generated by collision with objects or interaction with virtual entities in manipulation training. Discrete asynchronous events are produced by means of pattern recognition from synchronous data streams like the identification of throwing or catching of balls in a juggling training system.

The representation adopted for describing these entities is the one of tables, which is mostly effective when the entities in the tables are all of the same type. Each table, in addition to the content, stores metadata information that allows contextualizing of the content. The labels are used for describing the table and the variables, like for expressing that data have been filtered, normalized, or transformed in a domain-specific way. Tables are also able to provide a set of operations that are specialized by the content following an object-oriented approach. For example, a table containing a function of time could be encoded using polynomials or a spectral basis, and then it could expose a function for producing the time-sampled data.

Tables can be organized in a hierarchical or relational way: the first type of organization is adopted when there is a clear hierarchy among the information elements, while the second gives more flexibility in the connections between the different tables. The underlying storage mechanism should also be able to support these two approaches.

Variable Types

The system specifies the semantics of the variables used in the tables and uses them for supporting the experimenter in analyzing data. The semantics is important for reuse and validation of the analysis. The variable types considered here are keys, conditions, sensor data, performance indices, and feedback. Each table, as in the relational database, has one or more keys that identify each entity (tuple in database terminology). Because of the presence of metadata in the table, each entity is identified by the combination of the metadata keys of the table and the keys of the entity: a sample can be identified by combining the identifier of the session and the time inside the session. Other types of keys correspond to indices to segmentation operations, like stroke in rowing, or frequency index when a table represents spectral data. Condition variables depend instead on the training protocol. Examples of them are the current target element, the difficulty level, and thresholds. In general they correspond to independent variables that are used as factors in the analysis. Performance indices are the values computed in real time based on participant performance based on sensor data and conditions, like error scores or distance to target. Finally, feedback variables express the information that is used for triggering the stimulation of the participant in response to his or her actions.

Workflow and Skill Schemas

Defining the contextual information requires a definition of a schema for describing and annotating the activities performed in the workflow. This is in line with some other domain-specific works for the definition of database schemas for representing experiments such as in neuroscience (Bradley et al., 2005; Buneman et al., 2004). In the system discussed here the context schema is built around the basic concepts of skill training using concepts such as task, group, participant, template participant, session, trial, and block. The Unified Modeling Language (UML) structure of these concepts is shown in Figure 10.4, and some of them need to be explained. The term *task* is used here to describe one part of the workflow cycle, and it corresponds to the execution of an experiment, or an evaluation session or training or capturing session. Another peculiar element of the representation is the *template participant* that is a virtual participant that is used through the system for representing the specification of the task performed by a participant in a group. The session associated to the template participant has no associated data streams but only contextual information containing setup information. In this way the designer specifies the characteristics of

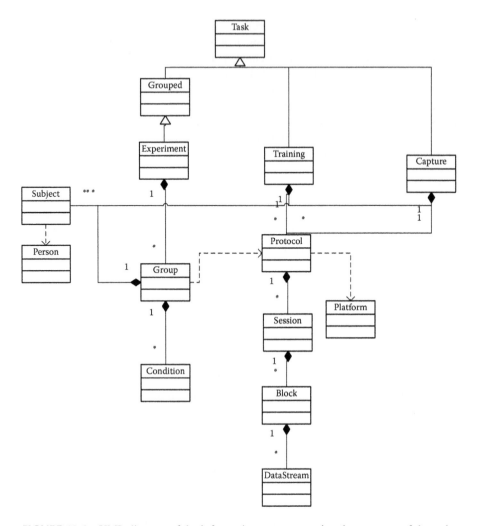

FIGURE 10.4 UML diagram of the information system covering the structure of the tasks in the workflow and the annotation of the recordings with performance indicators.

the template in every session as models and parameters, and then when a real participant from a group arrives, the corresponding template is instantiated. A mechanism of locking the template allows for keeping the protocol consistent.

A schema that parallels the one of the workflow is related to the representation of the domain skill and the curriculum schema. This schema should be able to cover:

1. Performance indicators of the given skill
2. Environment in which the skill has been measured, in particular for being able to perform measures of transfer of training from virtual to real world
3. Progress of the participant along the protocol in a form of curriculum

For meeting the above, the proposed schema contains the concept of measure that, for a given participant and given moment in time, in addition to the numerical value, is described by the platform in which has been taken and the definition of the measure as taken from a list of measures possible for a given domain. A challenging aspect of this schema is the unique definition of the measure because it can depend on a complex model or a number of highly contextual parameters. For this reason the designers should make the measure definition reproducible and clearly defined.

Data Processing

The type of processing performed in the off-line phase is common to that performed in experimental psychology: data are processed bottom-up, extracting statistics from participants and then synthesizing them for extracting statistically valid results. The data management framework discussed here provides two elements: a scheme for accessing and processing data, and the reuse of analysis and visualization blocks across experiments.

The accessing and processing of data are expressed by means of functional expressions aimed at avoiding the use of an explicit iteration loop, making it thus easier to be understood, to be manipulated with a GUI, and to be parallelized (Dean and Ghemawat, 2008). The adopted operations, applicable with a table containing data or visualization entities, are the following:

- Selection of elements based on criteria (e.g., specific training session or user)
- Application of functions over every element (e.g., scoring or classification)
- Aggregation of elements producing one or more results (e.g., computation of statistics or segmentations)
- Collection of elements from a table inside a table, a form of aggregation

The last operation is important when dealing with information structured hierarchically. For example, given a table of all the sessions from participants, a first step decomposes each session in segments (like strokes in rowing), then another computes statistics per strokes, and finally they are aggregated for every participant.

The functional representation of the operations over data allows also implementing a caching mechanism. The aim is to allow the user to easily take advantage of previously performed computations over data, an aspect that is important when the dataset is large or the operations are lengthy, like the training phase of many machine-learning algorithms.

Finally the reuse of blocks is provided by means of a dynamic type scheme over data. Each processing block can be applied over a table if the input value has the required variable element and specific labels in its metadata. It is also possible to meet block requirements by transforming the input to meet the requirement. For example, a rowing stroke can be segmented by means of the derivative of the horizontal angle of the oars; if the derivative is present it is used, otherwise it is computed. The execution of these transformations can be performed by means of the operations provided by the table or by general functions.

Skill-Processing Toolbox

Provided with the data structures and the processing framework presented above, it has been possible to create a library of operations that can be reused for training in VE. These operations are common to different skill domains. Following are examples, in increasing order of skill abstraction:

- *Trajectory segmentation and representation of segments*: Different segmentation methods are possible, but the type of operation over segments is common.
- *Trajectory comparison and variability measures*: This is a key element for comparing participant performance at different scales.
- *Coordination measures*: Phase indices for comparing actions performed by different elements, with them being different subjects or different limbs.
- *Aggregation of different measures for providing a single performance index*

The following section will present a specific application of the system to the domain of rowing, training in VE, with a customization of the toolbox.

ROWING CASE STUDY

The data management methodology discussed in the chapter has been applied in the context of the research activities on the SPRINT rowing training system (Ruffaldi et al., 2009). The SPRINT system is a multimodal VR environment built around a custom mechanical platform capable of capturing and displaying forces of rowing action. This platform is an interesting example for this discussion for several reasons:

1. Presents both evaluation of subjects and training
2. Exposes the problem of analysis of data
3. Deals with multimodal capturing and feedback from a variety of sensors

There are two environments in which research on SPRINT takes places, the VR setup and a boat. The real boat environment is used to capture subject performance in the real world, supporting the design of the virtual scenario. The instrumented boat allows the acquisition of several aspects, such as the kinematics of oars, seat position, and back angle, and the forces applied on the oar. For some analyses binaural data have been recorded as well. The setup of this work deals with 125 Hz sampling, although internally velocity is estimated at a higher rate. The VR environment is instead characterized by the mechanical platform that measures, at the same 125 Hz, the oar angles, seat position, and force applied, estimated by means of a model that correlates fan speed with applied force. In addition, body kinematics is acquired by means of a motion capture system, providing information not elsewhere available.

The key element of this scenario is the aim of evaluating subject performance for comparison against other subjects or for estimating the effectiveness of training both in real and virtual contexts. The additional step that can be employed with data is the synthesis of rower behavior for animating virtual humans that can be used in training protocols (Filippeschi et al., 2009).

The SPRINT system and the associated data management tool have been used in several workflows at different stages of development: three involved a simpler platform based on the Concept2, two were based on the full VE platform, and one involved the integration of boat data. The result is a total of six different protocols.

In the SPRINT system the architecture has been adopted first in the real-time modality for the analysis of subject performance and then in the off-line modality for the training of machine learning models during the design phase and then for the analysis in the analysis phase.

REAL-TIME MANAGEMENT

The real-time modality is based on a MATLAB® Simulink®* model producing results that are sent to a graphic rendering model based on the XVR framework. The Simulink model is different for every experiment and evaluation, but it is characterized by the same structure and it reuses many elements as taken from a common library. The actions of the model in line with the discussed architecture are as follows:

1. Acquisition of data from the sensors, integrating data from different devices like Vicon, and execution of specific filtering
2. Simulation of the boats dynamics, including the simulation of dynamics of the real participant and the virtual humans as part of the team or as competing boats
3. Extraction of the performance indicators of the real participant based on the above level
4. Execution of the training protocol as a state machine
5. Control of the feedback based on the result of the previous step

The performance analysis is performed at different time scales. The rowing stroke has been decomposed in phases by means of a classifier based on the neural network model trained over manually labeled data. In this way the system is able to provide events of phase change and subwindows associated to a whole phase (Johard et al., 2011). This phase decomposition corresponds to domain-specific gesture recognition. The phase information is also able to trigger events in the pedagogical engine, in particular in the context of technique analysis.

In the case of technique analysis the approach is to provide scores of performance based on the comparison with a model constructed from an expert. In one of the SPRINT scenarios, experts have been recorded while rowing at their best and then with specific errors. The whole dataset has been processed in the off-line phase for creating a model that is then used in real time for scoring the presence of error. Due to the fact that the error information should be provided to the user as near as possible to the moment of action, the recognition algorithm cannot be applied to the last stroke, but instead on a variable time window triggered by phase recognition.

For later analysis of the training execution, the system stores all the information produced, in particular the original raw sensors, the performance indicators, and the

---

* MATLAB® and Simulink are registered trademarks of The MathWorks, Inc.

feedback flags. This information, together with metadata that identifies the task and also the specific model being employed, allow evaluating in the offline phase both participant behavior and system performance.

OFF-LINE MANAGEMENT

In the off-line phase the data management system has been employed for the SPRINT system for accessing data from different experiments in a uniform way, covering six experiments plus six capturing sessions for a total 117 participants and 18 GB of recordings covering 143 hours of training and capturing. The common organization allowed for analysis code across the experiments to be reused at different levels, covering, for example, calibration per platform, stroke and phase segmentation, analysis of timing of phases, analysis of subject kinematics, and visualization of participant performance.

CONCLUSIONS AND PERSPECTIVES

The design and implementation of a data management scheme for training in VE presented several challenges, not only due to the variability in the characteristics of VE and training protocols but also in the fact that the final user of such a system should not be a programmer but instead a field expert. The organization of such a system has been presented in this chapter by taking a high-level view of the problem and presenting how skill and VE-specific aspects are covered by technical solutions. Due to the increasing number of VE setups and possible training experiments, there is a need for streamlining many of the management aspects, in particular for the reuse of developed code and adoption of best practices. There are two aspects that have not been covered by this chapter and that are an important element for future data management systems for VE: provenance and collaboration.

Provenance is the term used to identify the ability to track the flow of data from acquisition to publishing in journals and on Web sites. It is specifically related to the domain of research in which workflows are very flexible and variable, depending on methods and specific experiments. The purpose of provenance is to provide reproducible research, primarily for review purposes, allowing other researchers to test separately the results. As discussed by Mesirov (2010), there are two elements in a provenance system: an execution engine that processes data and a publishing environment for presenting the results associated to that. In addition to reproducing and reviewing, the provenance computational model has the advantage of supporting the caching of intermediate results, allowing the updating of results only when changes have been performed. In the future it is advisable to extend the system presented in this work toward provenance by adapting the data management system. It could be also possible to employ a versioning system for keeping track of the evolution of data entities, the computed models, and their use in training sessions (Ledlie et al., 2005). The other interesting aspect is the collaboration among designers and experimenters during the analysis process that should take advantage of the facilities of distributed computing and storage of the cloud.

A final comment is related to the extension of the methodology discussed here to other domains, in particular in the context of skill training in robotics, in which data management has similar characteristics and challenges.

ACKNOWLEDGMENTS

The work presented in this chapter was supported by the SKILLS integrated Project (IST-FP6 #035005, http://www.skills-ip.eu) funded by the European Commission.

REFERENCES

Arasu, A., Babu, S., and Widom, J. 2004. CQL: A language for continuous queries over streams and relations, *Database Programming Languages*, 123–124, Springer.

Bradley, D.C., Mascaro, M., and Santhakumar, S. 2005. A relational database for trial-based behavioral experiments. *Journal of Neuroscience Methods*, 1411, 75–82.

Buneman, P., Khanna, S., Tajima, K., and Tan, W.C. 2004. Archiving scientific data. *ACM Transactions on Database Systems TODS*, 291, 2–42.

Daly, L., and Brutzman, D. 2007. X3D: Extensible 3D graphics standard. *IEEE Signal Processing Magazine*, 24(6), 130–135.

Dean, J., and Ghemawat, S. 2008. MapReduce: Simplified data processing on large clusters. *Communications of the ACM*, 51(1), 107–113.

De Roure, D., and Goble, C. 2009. Software design for empowering scientists. *Software, IEEE*, 261, 88–95.

Dougherty, M.T., Folk, M.J., Zadok, E., Bernstein, H.J., Bernstein, F.C., Eliceiri, K.W., et al. 2009. Unifying biological image formats with HDF5. *Communications of the ACM*, 5210, 42–47.

Figueroa, P., Bischof, W.F., Boulanger, P., Hoover, H.J., and Taylor, R. 2008. Intml: A dataflow oriented development system for virtual reality applications. *Presence: Teleoperators and Virtual Environments*, 175, 492–511.

Filippeschi, A., Ruffaldi, E., Frisoli, A., Avizzano, C.A., Varlet, M., Marin, L., et al. 2009. Dynamic models of team rowing for a virtual environment rowing training system. *International Journal of Virtual Reality*, 4, 49–56.

Folk, M., Cheng, A., and McGrath, R.E. 1999. HDF5: A New File Format and I/O Library for Scientific Data Management. *Astronomical Data Analysis Software and Systems VIII Proceedings*, Astronomical Society of the Pacific, Mehringer, D.M., Plante, R.L., and Roberts, D.A. (eds.), 172.

Friedman, D., Brogni, A., Guger, C., Antley, A., Steed, A., and Slater, M. 2006. Sharing and analyzing data from presence experiments. *Presence: Teleoperators and Virtual Environments*, 155, 599–610.

Gerbaud, S., Mollet, N., Ganier, F., Arnaldi, B., and Tisseau, J. 2008. GVT: A platform to create virtual environments for procedural training. *Virtual Reality Conference, 2008. VR'08*. IEEE (pp. 225–232) Lin, M., Steed, A., Cruz-Neira, C. (eds.), Reno, NV.

Gosink, L., Shalf, J., Stockinger, K., Wu, K., and Bethel, W. 2006. Hdf5-fastquery: Accelerating complex queries on hdf datasets using fast bitmap indices. *Scientific and Statistical Database Management*, 2006. 18th International Conference on, pp. 149–158, Grossman, w. (ed.), Vienna, Austria.

Jenkins, J.M., Caldwell, D.A., Chandrasekaran, H., Twicken, J.D., Bryson, S.T., Quintana, E., et al. 2010. Overview of the Kepler science processing pipeline. *Astrophysical Journal Letters*, 713, L87.

Johard, L., Filippeschi, A., and Ruffaldi, E. 2011. Real-Time Error Detection for a Rowing Training System, *SKILLS 2011*, Montpellier, France.

Ledlie, J., Ng, C., and Holland, D.A. 2005. Provenance-aware sensor data storage. *Data Engineering Workshops, 2005. 21st International Conference on* (p. 1189) Aberer, K., Franklin, M., Nishio, S. (eds.), Tokyo, Japan.

Mesirov, J.P. 2010. Accessible reproducible research. *Science*, 3275964, 415.

Ruffaldi, E., Filippeschi, A., Frisoli, A., Avizzano, C.A., Bardy, B., Gopher, D., et al. 2009. SPRINT: A training system for skills transfer in rowing. In T. Gutierrez and E. Sanchez (eds.), *SKILLS09 International Conference on Multimodal Interfaces for Skills Transfer.* pp. 87–90. Bilbao, Spain.

Ruffaldi, E., Filippeschi, A., Avizzano, C.A., Bardy, B., Gopher, D., and Bergamasco, M. 2011. Feedback, affordances and accelerators for training sports in virtual environments. *MIT Presence*, 201.

Saal, L.H., Troein, C., Vallon-Christersson, J., Gruvberger, S., Borg, A., and Peterson, C. 2002. BioArray Software Environment BASE: A platform for comprehensive management and analysis of microarray data. *Genome Biol.*, 38, 1–3.

Sewell, C., Morris, D., Blevins, N., Barbagli, F., and Salisbury, K. 2004. An event-driven framework for the simulation of complex surgical procedures. *Medical Image Computing and Computer-Assisted Intervention—MICCAI* 2004, 346–354. Larsen, R., Nielson, M. (eds.) New York: Springer.

Section IV

SKILLS Demonstration Platforms

Section IV

11 Design and Evaluation of a Multimodal Virtual Reality Platform for Rowing Training

Emanuele Ruffaldi, Alessandro Filippeschi,
Manuel Varlet, Charles Hoffmann,
and Benoît G. Bardy

CONTENTS

INTRODUCTION

The advancements in virtual environment (VE) technologies, the improvements in models of motor control, and the availability of sophisticated data analysis techniques are all contributing to new practices in sport training. Classic training is now combined with new virtual reality (VR) training environments with specific feedback able to monitor performance, reduce training time, or allow training in places or moments in time not possible before. Rowing presents interesting research challenges and opportunities due to the combination of technique, strategy, and biomechanical elements, all contributing to performance. This chapter presents, from design to evaluation, the SPRINT research platform developed in the context of the SKILLS project. After introducing current research and training in VE, the SPRINT platform is presented. Then the results obtained in training specific skill accelerators on the platform are detailed.

SPORT TRAINING IN VIRTUAL REALITY

Assistive technologies have improved through the years, and today they contribute to increased physiological, biomechanical, or perception-action skills and provide coaches and athletes with fundamental information for shaping and programming training. Training in VR has received a great deal of attention in domains involving highly demanding cognitive tasks such as aeronautics or industrial maintenance (e.g., Aoki et al., 2007). The reduction of complexity of VR setups and the improvement in motion capture technology have allowed the extension of VR to physical training (Bailenson et al., 2008) and more interestingly for the present purpose to sport training, often using robotic and haptic systems (Multon et al., 2011). Research in sport training includes the investigation of complex skills in goal-oriented and task-oriented training situations (e.g., Wulf and Shea, 2002), as well as of specific perceptual or motor patterns. Bideau et al. (2003), Iskandar et al. (2008), and Vignais et al. (2010), for example, have developed simulated football scenarios for goal-keepers using immersive VR in order to evaluate how perception influences their decisions about when and how to move. In rowing, a similar approach has been proposed by Wolf et al. who recently, in parallel to this work, developed an immersive rowing system coupled with haptic feedback (von Zitzewitz et al., 2008). Rowing is an interesting sport in which VR training can be efficient due to the periodic and constrained nature of rowing action. Rowing requires athletes to be skilled in several areas to achieve a good level performance. Although it is difficult to rank them by order of importance, physical competencies are surely the most time-consuming skills to be developed, and rowing is often integrated with running and weight-lifting sessions. Skills such as rowing in team or mastering technique are seldom trained in a systematic way and are not continuously monitored. Coaches typically give advice about technique and team coordination verbally during training, with rare in-depth quantitative or qualitative analyses of the perceptual-motor behaviors involved. In sum, existing rowing training systems favor physical capabilities development over techniques or perceptual motor variables. The SPRINT system described in this work has been developed to train technique-based, perceptual, and physiological aspects of rowing using VR.

An initial task analysis for rowing (see Nolte, 2005, for details) led us to envisage three main training areas: technique optimization, energy management, and team coordination. Each area is composed of specific tasks (e.g., execute an efficient stroke, stabilize energy consumption, synchronize with other crew members), which are segmented in several subtasks (e.g., push with the legs, increase power output, apply more force at the inflexion point, etc.).

THE SPRINT PLATFORM

Design

SPRINT was designed to meet training needs, searching for the best compromise between variables useful for training and system complexity, and providing rowers with the same degrees of freedom and the same rigging settings they encounter in outdoor rowing. However, some boat degrees of freedom are not included, and force rendering is simplified in order to limit hardware complexity and portability. Similar decisions have been taken with software development, and features not directly involved in training were not introduced. The adopted design is highly modular, allowing future enhancement of existing parts and implementation of new parts at both the mechanical and software levels.

System

Trainees using SPRINT (see Figure 11.1) are in the center of a training loop, rowing on the mechanical platform, which is the main interface. Performance is captured by a set of sensors and analyzed via software in order to feed back information to the users, thus closing the loop. SPRINT is composed of four components—mechanics,

FIGURE 11.1   *(See color insert.)* The SPRINT system.

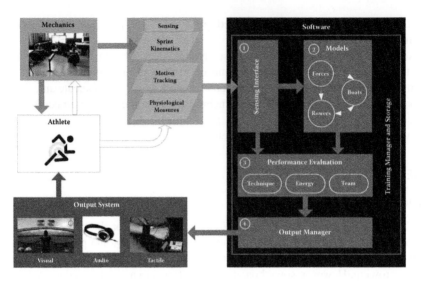

FIGURE 11.2 *(See color insert.)* Architecture of the SPRINT rowing system.

sensing, software, and multimodal output—which are described below. Figure 11.2 shows the overall architecture and the various components.

The *mechanical system* is composed of a steady rail and two steady boxes. The rail bears a sliding seat and a foot-stretcher. Height of the rail and position of the foot-stretcher can be adjusted. The boxes support oars and force rendering devices. They can be used together or one at a time, allowing sculling and sweep rowing, with possible regulation of geometry and load. Force rendering is provided by a fan mounted on a flywheel, making load dependent on the oar's angular speed, angular acceleration, and airflow blown by the fan (but not handle height).

The *sensing system* varies according to the training purpose. The simplest configuration is composed of six encoders measuring oars' angles and fan angular speed, plus an infrared sensor capturing seat displacement. This basic configuration can be enhanced with motion tracking systems (e.g., Vicon by Oxford Metrics, United Kingdom) or physiological recorders (e.g., Cosmed K4 and Polar belt), if necessary.

The *software system*, developed in Simulink (Mathworks, Natick, Massachusetts), is the core of the SPRINT platform. It is composed of four levels. A first level contains the interfaces with the sensing devices. The second level implements the physical models for force rendering, athlete, and boat-oars-rower systems. These models provide an estimation of user performance in terms of boat motion, rowers' internal forces, torques, inertia, and energy. The third level indicates performance. According to the raw data and the physical models, performance indices are computed for three skill elements: rowing technique, energy management, and team coordination. The fourth level includes the output manager, selecting the information displayed to the user through the SPRINT modalities. Visual, audio, and tactile outputs are decided according to the feedback selected for training. At this level, performance is mapped to feedback triggering, for example, switching the color of an element of the virtual environment, or delivering or not a feedback. In parallel to these four levels the SPRINT software organizes the various training sessions and ensures data storage. The user is able to set, automatically from a template or manually, the various training variables such as workout, rest times, distance to be covered, and feedbacks to be given. The relevant variables recorded in each training session are stored in a SQL database structure managing HDF5 datasets (see also Chapter 10).

The output system is composed of an immersive graphical environment, vibrotactile effectors, and an audio engine. The graphical component provides the scenario where the training takes place. The scenario is composed of a channel in which one or several boats move propelled by virtual rowers, which may be displayed with their movements precisely controlled. Training information in the graphic scenario may be conveyed by elements of the scenario (see the coordination experiment below), by symbolic elements immersed in the scenario (such as the arrow used in the energy management experiment below), or by elements not belonging to the scenario (such as the number superimposed on the scenario indicating the current pace). Vibrotactile displays are composed of vibrating motors housed in wristbands or belts worn by the user, delivering tactile feedback about performance, timing, or path motion (Ruffaldi et al., 2009, see also Chapter 8). Speakers provide audio information, delivered in a feedback form, or used to increase the scenario realism. Binaural recording of outdoor rowing allowed current user performance to be synchronized with the typical sounds of outdoor rowing.

ORCHESTRATION

A key characteristic of the output system is a messaging mechanism controlling most of the elements in the environment. The graphical application is a player of commands that can be used to display real-time performance, to deliver various stimuli for training purposes, or to playback from previous training sessions. Some commands allow reconfiguring the environment, specifying, for example, the number of boats and rowers. Others provide real-time updates of the environment. Some commands cover general aspects of the virtual environment, such as virtual time or camera position, while others control the placement and visibility of virtual boats and avatars. More specific commands display information related to a given training protocol.

TRAINING MODEL

Training tasks were selected according to task analysis, and accelerators—combinations of variables to be tracked, feedback, and protocols (see Chapter 2)—were selected in three steps. The taxonomy of VE training feedbacks and accelerators for sport training in general and for rowing in particular can be found in Ruffaldi et al. (2011). First, variables to be tracked were selected. Then selected feedbacks were chosen according to the literature on motor skills learning and training in VE (see Chapter 3). The accelerators included in this work are (1) a multimodal audio/vibro-tactile feedback about the rowing technique; (2) visual information in the form of an opponent avatar, about the management of energy stock during the race; and

(3) visual information about between-rower synchronization during team rowing. All information was presented on line modulated in real time by the rowing action during training protocols involving pre- and post-tests, learning and retention sessions. The results are presented below.

PROTOCOL DESIGN

Training protocols are structured according to a time hierarchy: A protocol is composed of days, sessions, and blocks, whose arrangement depends on the task, the selected feedback, fitness, and expertise of the rower. Task and fitness limit upper and lower bounds of blocks duration, rest time between blocks, and number of blocks per session (e.g., race simulation). For example, intermediate rowers may require longer training sessions than novices to acquire a new technique because they may have to annihilate their representation of the task or their already stable sensorimotor repertoire (e.g., Faugloire et al., 2009). However, long protocols can be carried out only by fit rowers; hence the final protocol should take into account both fitness and expertise. In the following sections, we report the results of three experiments recently performed on the SPRINT system, designed to evaluate the efficiency of the platform in the training of technique optimization, energy management, and team rowing.

LEARNING TECHNIQUE OPTIMIZATION

OBJECTIVE

Technique optimization training aims at providing novice rowers with an overall correct representation of the rowing cycle, allowing them to start rowing appropriately. Technique evaluation uses expert performances that were first recorded on the SPRINT platform. The obtained dataset was used to develop digital models of the technique features characterizing both correct rowing and technique faults. Some of these features were taken from literature (Nolte, 2005) and mapped on the variables available in SPRINT (e.g., deep blade entry maps on oars angles). Other features (e.g., timing of body limbs) were obtained from the recorded data. The feedback selection was done according to the training model.

EVALUATION

The evaluation followed a pretest, training, post-test, and retention design. The goal for the participants was to synchronize motion onsets of their legs, back, and arms during the drive phase. The information exchange relied on audio guidance, vibrotactile feedback, and delayed offline knowledge of results (KR). The participants were eight naïve rowers screened for handedness and general health. They were asked to row following the imposed timing. The timing pattern was captured from expert performances but with two simplifications: the load was removed in the first training part and the threshold on arm motion onset was loosened in order to ease the participant to fix at least arm motion. Participants were divided into two groups (four participants in each group) with or without vibrotactile feedback. The audio cue, available for all participants, consisted of two signals triggered at specified moments in time. These signals represented instants at which participants were instructed to start swinging their backs and bending their arms, respectively. The vibrotactile feedback was provided by two vibrating motors arrays: one housed in a wristband and the other embedded in a belt worn on the midriff. Vibrations on the wrist or on the back were triggered when the onset of arm or back motion exceeded a set threshold. The delay between vibration and audio allowed the participants to establish whether their movements were synchronized with the audio tone. All participants received a delayed KR about the missed cycles ratio over the total number of strokes after each trial.

Procedure: The experiment was carried out in three consecutive days: two days of training and a third day of test. The last day included a retention test, which was carried out with full load.

The onset times of arms and back motions were used as metrics compared to the trained reference. The back and arm timing errors e_a and e_b were computed for every stroke and made relative to the reference.

Results: Figure 11.3 shows the error for each pre, post, and retention session in the full-load condition after having averaged the error per session for compensating the different lengths in number of strokes. Both vibration and knowledge of results (VIB-KR) and KR participants generally

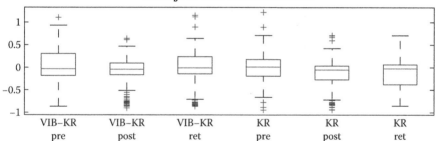

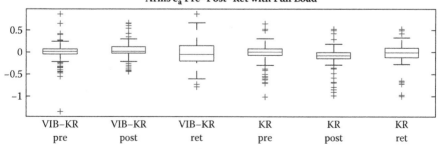

FIGURE 11.3 Statistics of the technique optimization experiment focused on the full load condition. In the top part the back timing error is presented for three sessions—pre, post, and retention—separated by the two groups, the experimental (VIB-KR) and the control (VIB-KR). In the top part the arm timing error is presented.

reduced their error of back and arms. The arm error was generally lower than that of the back. The vibrotactile feedback did not add any benefit when coupled with audio guidance and KR. Investigating the progress of subjects during training, they were not able to avoid errors most of the time, and they tried either to focus on one limb or to keep the right time lapse between limbs onset. These results indicate that stimulating the training of a multilimb activation pattern is challenging because it is perceived as a multigoal task, and, although the feedback scheme is promising, it has to be supported by a more focused protocol.

ACCELERATING THE MANAGEMENT OF ENERGETIC RESOURCES

OBJECTIVE

Olympic rowing events are conducted over a 2000 m race. Individual races on this distance last between 320 and 500 s. Rowing performance is constrained by several factors that should be taken into account during training, such as the rower's fitness status, the specific technique, as well as intra- and interindividual coordination. In the study described below, we focused on one other important factor, the ability of rowers to manage their energy stock during a 2000 m race. Garland (2005) reported that elite rowers adopt a particular pacing strategy. Their velocity corresponds to a fast-start profile, with the first 500 m performed at 103.3% of the average whole race speed, and with the subsequent sectors rowed at 99%, 98.3%, and 99.7% of the average speed, respectively.

Here we used virtual reality in order to determine if novice rowers were able to acquire and maintain this energy management skill during a 2000 m race, with positive consequences for rowing performance. We used an avatar on a screen located in front of the rowers to impose boat speed, in intrinsic units (i.e., in proportion of the actual capacity of the participants). We expected after training a better management of energy consumption for the avatar group compared to the control group. The protocol (Figure 11.4) was performed on the lightweight rowing platform based on the Concept2 indoor rowing machine by two groups of novice males. After a pretest both groups performed a 2000 m race twice a week, for a total of eight learning sessions. One group followed a classic indoor rowing training. The other group benefited from an energy-management information represented by an avatar boat visible on a large screen located in front of the participants. Participants (of the avatar group) were instructed to track the virtual boat, whose velocity was previously calibrated to follow the appropriate to-be-learned velocity profile along the 2000 m race. The virtual opponent was gradually removed at the end of the race along the 4 weeks of learning. After the 4 weeks training period, both groups achieved a post-test to evaluate the effect of the velocity profile accelerator. A retention test was performed 30 days later in order to evaluate the durability of learning.

A general analysis of variance (ANOVA) revealed a significant decrease in race duration between the pre- and post-tests. It also revealed an interaction effect

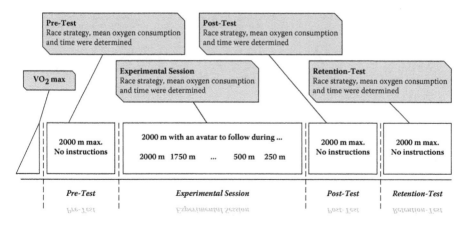

FIGURE 11.4   The energy management protocol.

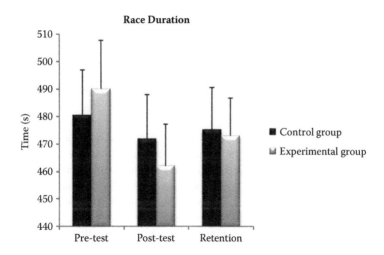

FIGURE 11.5   Race duration comparing avatar group (black) and control group (gray) for pre-, post-, and retention tests.

between groups and tests showing that the pre-post and the preretention difference concerned only the avatar group. Figure 11.5 summarizes graphically these results.

In addition we observed that the avatar group learned the expert profile and maintained it during the retention test (Figure 11.6). Concerning the oxygen consumption, we found an increase in VO_2 for the avatar group, which can be correlated with the fact that participants increased their rowing frequency compared to the control group. The control group did not reveal any differences between pre, post, and retention tests.

Our results indicate that virtual reality can be used to accelerate the learning of energy-related skills in a relatively short period of time (4 to 5 weeks). This learning can lead to a better performance in terms of race duration. These results open new

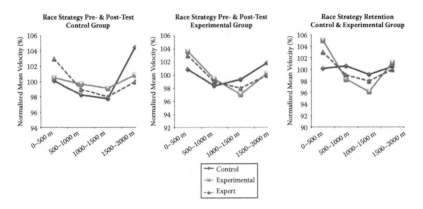

FIGURE 11.6 Race pace profiles comparing pre-test (straight black line), post-test (gray line), and the expert profile (dashed black line) for the control group (a) and the avatar group (b). Race pace profiles comparing control-group (straight black line), experimental group (grey line), and the expert profiles (dashed black line) in the retention.

issues concerning the transfer of pacing strategy to other sports including races of similar duration (6 to 8 minutes) such as running or sprint cycling, for instance.

LEARNING TEAM COORDINATION USING A VIRTUAL PARTNER

Objective

As mentioned above, performance in rowing depends on several factors such as the fitness status of the rowers, their energy management, and their technique (Baudouin and Hawkins, 2002). However, individual skills of rowers are often equivalent, and the difference in performance between two teams strongly depends on the ability of the athletes to row together in a highly synchronized way during the race (Hill, 2002). Although the synchronization between the movements of rowers is a significant factor of performance, the learning of team rowing coordination is limited in classical training situations. It is difficult for the coach to have an accurate estimation of the coordination of the team and to give efficient feedback to the trainees. Even if more accurate estimation of the synchronization is possible with video analyses, it does not allow the delivery of feedback in real time. Here we report a recent study (Varlet et al., submitted) in which we used virtual reality and real-time human movement capture technologies in order to accelerate the learning of team rowing coordination in VR (i.e., Filippeschi et al., 2009). We expected better team rowing coordination after training in VR, which would be stronger for the group that benefited from the real-time feedback.

Sixteen participants have been evaluated on the lightweight platform composed of a Concept2 indoor rowing machine located in front of a large screen (see Figure 11.7). The movements of the handle and the seat of the rowing machine were captured at a sampling rate of 100 Hz by using infrared marker based Vicon MX13 cameras. The participants performed pre-, post-, and retention tests in which they had to row as

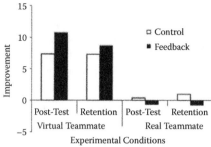

FIGURE 11.7 *(See color insert.)* Left: Experimental setup. (A) Coordination with real team-mate. (B) Coordination with virtual teammate. Right: Improvement compared to the pre-test of the synchronization for control (white) and feedback (black) groups in the different experimental conditions.

synchronized as possible with a virtual teammate displayed on a monitor in front of them, and in a transfer task with a real teammate (see Figure 11.7). After the pre-test, participants performed four learning sessions (one per day with a day break halfway) before performing the post- and retention tests, 1 and 4 days after the last learning session, respectively. The duration of the trials was 90 s, and we used the frequencies of 18, 24, and 30 strokes/min. Participants performed two trials of each frequency for pre-, post-, and retention tests, and four trials for each frequency for the learning sessions. All trials were counterbalanced.

Two groups of eight participants (mean age of 21.4 years) were composed. During the learning sessions, they were instructed to synchronize their movements with the virtual rower while having a real-time visual feedback giving either information about the coordination (Feedback group) or no information at all (Control group). For the Feedback group, we computed in real time the degree of synchronization of the trainee and used it to change continuously the color of the virtual teammate between red (not synchronized) and green (well synchronized). To measure the synchronization between the movements of the participants and their teammate (real or virtual), we computed the cross-spectral coherence giving an index of synchronization between 0 and 1 with 1 indicating a perfect synchronization and 0 indicating no synchronization (Schmidt and O'Brien, 1997). We averaged the synchronization measures at the level of the seat and the handle, and we computed then the percentage of improvement in post- and retention tests compared to pre-tests. The results are presented in Figure 11.7.

EVALUATION

A $2 \times 2 \times 2 \times 3$ repeated-measures ANOVA with variables of Group (Control and Feedback), Teammate (Virtual and Real), Test (Post- and Retention tests), and Frequency (18, 24, and 30 strokes/min) performed on the improvement percentages of participants yielded a significant main effect of Teammate ($F (1,7) = 57.05$, $p < .05$) and a significant interaction between Teammate and Group ($F (1,7) = 7.21$,

$p < .05$). As depicted in Figure 11.7, these results demonstrate that participants of the Control and Feedback groups produced better performance in post- and retention tests compared to the pre-tests while synchronizing with the virtual teammate, and that this improvement was stronger for the Feedback group that had real-time information about the coordination during learning sessions.

As expected, these results show that the learning of team rowing coordination is possible using VR technology and that the use of feedback giving information about the synchrony allows accelerating the learning in line with previous research on daily postural activity (e.g., Faugloire et al., 2005). However, our results did not reveal a significant influence of the learning sessions in VR on the participants' performance in the transfer task where they had to synchronize with a real teammate. Generally, our study showed the interest of VR and motion capture technologies for learning coordination in rowing, and encourages further exploration in order to develop learning protocols in rowing and other sport activities.

PERSPECTIVES AND CONCLUSIONS

The three training scenarios described above show several opportunities for research in the domain of VR and sport. First, at a methodological level, it is clear that it is possible to integrate and customize an expert performance profile inside training protocols, allowing the training of complex perceptuomotor elements such as the combination of activation patterns found in rowing. These profiles, as shown by the three accelerators, emerge at multiple levels of rowing behavior and cover different aspects of the task, from single stroke to higher-level strategy. The key element behind the profile identification, which has not been discussed in depth here, is the construction of a digital representation of the rowing skill that allows characterizing expertise and performance. This representation is a combination of existing knowledge and machine learning–based analyses that provide a working mechanism for scoring performance and controlling virtual participants.

The second fundamental aspect is the identification of training feedback and their combination with training protocols capable of maximizing the real-time analysis and synthesis capabilities of VR technologies. The embodiment of training in the form of a virtual human is the element that characterizes the accelerators presented above. In two cases the virtual human conveys the training feedback in the form of a partner, in one case for keeping a certain rowing distance, in the other for team synchronization. In both cases the virtual human is not an opponent that has to be fought, although this could be an added interest for increasing motivation and entertainment.

Two related aspects could be considered as sources for future investigation, feedback adaptation and protocol duration. The intensity and type of feedback provided to the user were fixed or partially progressive in our case, allowing for reduction of dependency. Although the results are encouraging, the adoption of an adaptive scheme of feedback would possibly increase the training effect, adapting the level of difficulty during the protocol. As a conclusion, we think that the approach presented in this work paves the way to more effective training strategies and experimental designs not only in the domain of rowing training but in VR sport training in general.

ACKNOWLEDGMENTS

The work presented in this chapter was supported by the SKILLS integrated Project (IST-FP6 #035005, http://www.skills-ip.eu) funded by the European Commission.

REFERENCES

Aoki, H., Oman, C.M., and Natapoff, A. 2007. Virtual-reality-based 3D navigation training for emergency egress from spacecraft. *Aerospace Medical Association* 78: 774–783.

Bailenson, J., Patel, K., Nielsen, A., Bajscy, R., Jung, S.-H., and Kurillo, G. 2008. The effect of interactivity on learning physical actions in virtual reality. *Media Psychology* 11: 354–376.

Baudouin, A., and Hawkins, D. 2002. A biomechanical review of factors affecting rowing performance. *British Journal of Sports Medicine* 36: 396–402.

Bideau, B., Kulpa, R., Ménardais, S., Fradet, L., Multon, F., Delamarche, P., et al. 2003. Real handball goalkeeper vs. virtual handball thrower. *Presence: Teleoperators and Virtual Environments* 12: 411–421.

Faugloire, E., Bardy, B.G., Merhi, O., and Stoffregen, T.A. 2005. Exploring coordination dynamics of the postural system with real-time visual feedback. *Neuroscience Letters* 374: 136–141.

Faugloire, E., Bardy, B.G., and Stoffregen, T.A. 2006. The dynamics of learning new postural patterns. Influence on pre-existing spontaneous behaviors. *Journal of Motor Behavior* 38: 299–312.

Faugloire, E., Bardy, B.G., and Stoffregen, T.A. 2009. (De)Stabilization of required and spontaneous postural dynamics with learning. *Journal of Experimental Psychology: Human Perception and Performance*, 35, 170–187.

Filippeschi, A., Ruffaldi, E., Frisoli, A., Avizzano, C.A., Varlet, M., Marin, L., Lagarde, J., Bardy, B.G., and Bergamasco, M. 2009. Dynamic models of team rowing for a virtual environment rowing training system. *International Journal of Virtual Reality* 4: 19–26.

Garland, S. 2005. An analysis of the pacing strategy adopted by elite competitors in 2000 m rowing. *British Medical Journal* 39: 39–42.

Hill, H. 2002. Dynamics of coordination within elite rowing crews: Evidence from force pattern analysis. *Journal of Sports Sciences* 20: 101–117.

Iskandar, P., Hanum, Y., Gilbert, L., and Wills, G. 2008. Reducing latency when using Virtual Reality for teaching in sport. *International Symposium on Information Technology.* In *Information Technology*, 2008. ITSim 2008. International Symposium on, Kuala Lumpur, Malaysia, 26–28 Aug 2008, 1–5.

Multon, F., Kulpa, R., and Bideau, B. 2011. Special issue: Virtual reality and sports guest editors' introduction. *Presence: Teleoperators and Virtual Environments*, 20(1).

Nolte, V. 2005. *Rowing Faster.* Champaign, IL: Human Kinetics.

Ruffaldi, E., Filippeschi, A., Frisoli, A., Sandoval, O., Avizzano, C.A., and Bergamasco, M. 2009. Vibrotactile perception assessment for a rowing training system. *Proceedings of the Third IEEE Joint Conference on Haptics 2009*, Salt Lake City, UT.

Ruffaldi, E., Filippeschi, A., Avizzano, C.A., Bardy, B.G., Gopher, D., and Bergamasco, M. 2011. Feedback, affordances and accelerators for training sports in virtual environments. *Presence: Teleoperators and Virtual Environments* 20(1): 33–46. Jones, L., Harders, M., Yokokohji, Y. (eds.), pp. 350–355.

Schmidt, R.C., and O'Brien, B. 1997. Evaluating the dynamics of unintended interpersonal coordination. *Ecological Psychology* 9: 189–206.

Varlet, M., Filippeschi, A., Ben-sadoun, G., Ratto, M., Marin, L., Ruffaldi, E., and Bardy, B.G. Learning team rowing coordination using virtual reality. *SPRINT.* Manuscript submitted for publication.

Vignais, N., Kulpa, R., Craig, C., and Bideau, B. 2010. Virtual thrower vs. real goalkeeper: Influence of different visual conditions on performance. *Presence: Teleoperators and Virtual Environments* 19: 281–290.

Wulf, G., and Shea, C.H. 2002. Principles derived from the study of simple skills do not generalize to complex skills learning. *Psychomonic Bulletin and Review* 9(2): 185–211.

Zitzewitz, J. von, Wolf, P., Novakovic, V., Wellner, M., Rauter, G., Brunschweiler, A., et al. 2008. A real-time rowing simulator with multi-modal feedback. *Sports Technology* 1: 257–266.

12 Training to Juggle with a Light Weight Juggler (LWJ)

Julien Lagarde, Daniel Gopher,
Carlo Alberto Avizzano, Vered Erev,
Vittorio Lippi, and Gregory Zelic

CONTENTS

INTRODUCTION

The juggling skill is the mastery of performing juggling tricks. A trick requires performers to execute and monitor the concurrent tossing and catching of multiple objects (e.g., balls, cones) in consecutive cycles, which follow a predetermined pattern. These patterns, labeled tricks, vary in their spatial and sequential format and in the number of concurrent objects used. Three-ball cascade, five-ball cascade, three-ball snake, and four-ball shower are some examples of such tricks. Competent juggling entails the ability to perform a trick repeatedly, rhythmically, and accurately without dropping balls, as well as mastering an increasing number of different tricks and switching between them. As such, the juggling task provides both the learner and the researcher a rich environment for examining and exploring different aspects of skill acquisition.

The science of juggling has brought significant insight to the understanding of complex cognitive and perceptual-motor skills. Juggling represents an emblematic case of goal-directed behavior under severe physical constraints, and as such has attracted attention in several fields, such as mathematics, motor control, and neuroscience (Beek and Lewbel, 1995; Draganski et al., 2004). Even in one of the easiest tricks, the three-ball cascade, the acquisition of juggling skills requires a great deal of practice (Huys et al., 2004), imposes clearly drastic changes in the repertoire of coordination between vision and arm movements (Beek and van Santvoord, 1992; Draganski et al., 2004; Huys et al., 2004), and is characterized by a high degree of spatial-temporal behavioral organization (Post et al., 2000). This pattern of whole-body movements is essentially captured by relative timing, and is readily measured by frequency and phase relations between hands, arms, eyes, and postural movements (Huys et al., 2004).

In this chapter, we report the results of three evaluation studies performed with a Light Weight Juggler (LWJ), a new virtual reality juggling system developed within the framework of the European SKILLS project to train novice jugglers.

The basic design of the LWJ and the accompanying training strategies were driven by the findings reported above. Accordingly we focused on the invariant relative timing pervasive in juggling at the expense of other properties present in the real environment, notably kinetic rendering of the interaction with the balls. We assumed that this choice would maximize the transfer of training to juggling with real balls while minimizing the technological cost of the development of the training platform.

Achieving a juggling pattern is a complex rhythmic task that requires moving the hand to catch and toss balls in order to keep a stationary phasing among the balls. A good performance in juggling imposes continuously adjusting the timing of the arms movement. In juggling cascade patterns, there are three critical time components: (1) the time an object is in the air, (2) the time an object is in the hand, and (3) the time that the hand is empty and moving (that is, the between a catch and a throw) (Beek and Turvey, 1992). The relationship between these variables was expressed by Claude Shannon in the formula $(F + D) \times H = (V + D) \times N$, where F denotes the time a ball is in the air, D denotes the time a ball is in a hand, V denotes the time a hand is vacant, N denotes the number of items juggled, and H denotes the number of hands (Beek and Lewbel, 1995). An important ratio that can be deduced from this equation is the ratio of object-in-hand time to hand-cycle time, called the *dwell ratio* and represented by the letter K. As the K value increases, the chance of a ball collision decreases because the average number of balls in the air is reduced (Haibach et al., 2004). A stabilized juggling pattern is often described as having K value of 0.75, whereas expert juggling can be characterized by the ability to operate at different K value ranges to restore stability at the 0.75 level (Beek, 1992; Beek and Turvey, 1992; Beek and Lewbel, 1995). Note that the 0.75 K value can be obtained by an infinite number of combinations of time in the air (extracted from the height and width of a ball spatial trajectory) and the time that the ball is held in the hand before tossing it. For example, tossing balls higher (i.e., resulting in a longer duration in the air) will also require holding them longer in hand after a catch before tossing them, and vice versa. An expert juggler can automatically vary his or her responses to accommodate for such changes (e.g., Beek, 1992).

DESCRIPTION OF THE LWJ HARDWARE, SOFTWARE, CAPTURING, AND ANALYSIS

The LWJ hardware consists of a Polhemus Liberty tracker, a personal computer, and an LCD screen (see Figure 12.1). In line of principle, the LWJ requires a tracking

FIGURE 12.1 *(See color insert.)* A view of the Light Weight Juggler (LWJ).

system for user's hands and a display to show the virtual environment. Some variations are possible; in particular the system is ready to be used with a stereoscopic visualization system that will allow a better position control in space. A user's hands are tracked through the Polhemus Liberty 3D tracker. It is used to get the hand position in real time. It records the position and the orientation (not used in this application) of two sensors held by the user. The application is controlled by the user just through this device: ball tossing is triggered by hand acceleration and ball speed just after the toss is set to hand speed.

Visualization system: The basic visualization hardware consists of a monitor displaying the virtual environment showed to the trainee. A second monitor may be used as an interface to experiment parameters and results. The system can be easily adapted to produce a stereoscopic updated view according to the user's head.

Computing unit: Running the LWJ requires reading the sensors, updating the virtual environment state, and rendering the scene. The hardware currently used is a personal computer.

Software: The LWJ software is made up of two main modules taking care, respectively, of rendering the virtual reality, reading the sensors, and simulating ball dynamics (and interactions) accordingly. The virtual environment is implemented with XVR, and the sensor reading and the virtual environment simulation are implemented with MATLAB® and Simulink®.*
The two parts work as two separate processes and communicate through

---

* MATLAB® and Simulink® are registered trademarks of The MathWorks, Inc.

User Datagram Protocol (UDP). A three-dimensional representation of juggling portrays two hands controlled in position by the user and three balls to be juggled. This visual interface is programmed with XVR. This is purely a graphic rendering receiving object positions from a MATLAB/ Simulink scheme that reads the sensors and simulates the physics.

Environment simulation: Trajectories are numerically integrated; tossing and catching are managed by finite state machines that represent the state of hands and balls within the exercise.

Simulated environment main parameters: Various parameters can be set up according to the experimental needs. The most important is the speed that is used to tune the difficulty of the task. The speed consists in a factor that is multiplied to the integration time of the whole simulation affecting both speeds and acceleration. Changing the speed produces a fast/slow motion effect in the virtual environment. The gravity acceleration and the thresholds in speed and acceleration for tossing can be tuned as well to achieve a feeling of realism and comfort according to the other settings.

Sound and vibrotactile stimulation: The system may take into account different kinds of feedback besides the visual rendering. A vibrotactile and an audio feedback have been introduced to test various combinations of touch and sound on the users. These outputs are controlled directly by the MATLAB/ Simulink application.

Analysis tools: Several analysis tools have been developed on the LWJ platform for a particular experiment or with demonstrative purpose. The main performance index is the number of missed balls compared with the number of tosses performed. This index is used because it is immediate to compare it with the performance in the real world and has been a feature of all existing systems. This simple index has been extended, adding some goals in terms of successful toss/catch sequences to be performed without interruption. Also the phase difference between hand and between ball positions has been calculated and output as an online analysis tool.

AUDIOTACTILE PERIODIC EVENTS IMPROVE JUGGLING COORDINATION DYNAMICS

In order to improve the training within the LWJ, we took inspiration from the discovery of effective stabilization of coordination. These previous results were obtained for various examples of interlimb and limb-environment coordination dynamics, in particular in learning a new pattern of bimanual relative phase (Tuller and Kelso, 1989; Zanone and Kelso, 1992). This approach has also been investigated for multimodal coordination, engaging movement in relation to audiotactile periodic stimuli (Lagarde and Kelso, 2006). Other results have been obtained in the case of coordination involving the *polyrhythms* (Kelso and DeGuzman, 1988; Peper et al., 1995)— that is, the components, being limb movements or external periodic events evolve at different frequencies and are frequency locked with particular ratios. These ratios are well predicted by formal models of coupled oscillators (DeGuzman and Kelso, 1991).

Based on a previous study on intermediate jugglers with real balls (Zelic et al., 2012) we assumed that audio-tactile periodic pacing could improve the acquisition of the three-ball cascade. We called this solution a *multimodal rhythm trainer*. Juggling skills are characterized by the assembly of multiple subskills, including coordination with multifrequency relations, mainly among the eyes looking at the ball pattern in the air and the hand (Huys et al., 2004). Based on a previous study on the stabilization of juggling coordination of intermediate-level jugglers by several audiotactile metronomes (e.g., Zelic et al., 2012) we implemented a rhythm trainer consisting of a multifrequency sensory stimulation. The period of the sound was scaled to match the period of one cycle of one ball, while the period of the tactile pacing matched the period of one cycle of the hand. The tactile pacing was presented at the left and right wrists in a syncopated order (anti-phase), corresponding to the relative timing between the hands in the three-ball cascade. We implemented and tested in the LWJ two very distinct types of rhythm trainer: a closed-loop version and an open-loop version. The open-loop version consisted of presenting audiotactile stimulations with fixed periods, updated only from one trial to the next based on the period of the trainee. In the closed-loop version the audiotactile stimulations were timed according to the current behavior of the trainee: tactile vibrations were provided at the hands each time a toss was achieved, and a beep was provided every three vibrations to maintain the main three-ball cascade frequency ratio.

We recruited 15 adults from Montpellier University having no prior juggling experience; they were paid for their participation. The participants practiced the three-ball cascade during 2 weeks for 10 sessions including pre-test, and nine training sessions, and then passed a post-test session 1 week later. Five participants were assigned to each of the groups and performed nine learning sessions. The control group consisted of those training with real balls. Each session lasted 20 minutes without supplementary instructions. Each session finished with the 5-minute test during which participants had to perform as many consecutive catches as possible, and the number of consecutive catches was recorded. The transfer of learning was evaluated during this session with real balls, participants were instructed to juggle as many consecutive catches as possible, and the number of consecutive catches was recorded.

The transfer of training to real balls seemed more efficient in the closed-loop condition than in the open-loop condition (Figure 12.2). Like in the control group, four subjects in the open-loop condition reached a performance of at least 10 consecutive catches, while in the closed-loop condition only one subject reached this level. However, the statistical analysis (ANOVA [analysis of variance]) did not confirm significant differences between the three groups, probably due to the small number of trainees in each group. We hence found an indication that a rhythmic trainer in the LWJ augmented with adapted audiotactile combinations can lead to the same level of performance after a learning phase than when training with real balls.

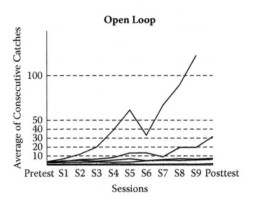

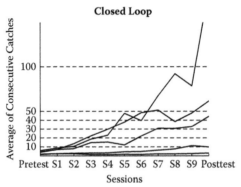

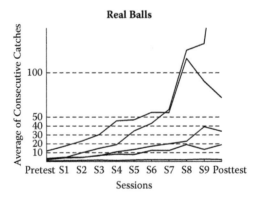

FIGURE 12.2 Comparison of juggling performance with real balls between the three groups of training. The average number of consecutive catches across trials is shown, for the five subjects of each group, recorded during the pre- and post-tests, and at the end of each of the nine training sessions.

COGNITIVE PERSPECTIVES ON TRAINING AND TRANSFER OF JUGGLING WITH THE LWJ

We also analyzed and examined the cognitive components of the juggling skill with an attempt to understand their role and contribution to learning and performance of juggling. In a cognitive science framework, skill is defined as a well-organized knowledge in memory (i.e., Anderson, 1992; Gopher, 1994; Logan, 1988). From this perspective juggling skill can be conceptualized as including three primary components:

1. Pattern memory: The ability to identify, represent, and memorize the pattern and sequence of objects in a trick
2. Acquisition of a stable representation of the spatial-temporal relationship of catching and throwing objects
3. Motor competency of tossing and catching juggled objects

While the first two have major cognitive constituents, motor competencies call for coping with gravity and force factors associated with the weight of balls, their tossing, catching, and moving between hands. These were absent in the LWJ training platform. Hence the aim of the line of studies was to investigate the development with training of the two other and more cognitive components of juggling (1 and 2 above) and examine the transfer of training from LWJ training to juggling with real balls. Two experimental studies were first conducted in the LWJ platform to investigate the value of cognitive accelerators for the acquisition of juggling skills. One manipulated the global speed of the virtual environment, the other experimented emphasis change of the ball in air/ball in hand dwell ratio. A third experiment examined the contribution of this training to the juggling of real balls. The results of these studies are briefly summarized here.

LWJ JUGGLING UNDER GLOBAL SPEED MANIPULATION

A key requirement in the ability to juggle a specific trick and distinguish it from others is the representation and memorization of the required sequential pattern of tossing and catching the juggled objects. For the novice juggler who needs under normal conditions to acquire the ability to coordinate tossing and catching of balls, using his two hands and maintaining a predetermined spatial pattern, order, and pace of objects, things happen "too fast." The immediate implication is that in real juggling beginner trainees have a hard time keeping track of the concurrent ongoing events. They do not have sufficient time to recognize and memorize the ball pattern, observe the consequences of their controlled movement, and encode and store all necessary information in memory. Another difficulty is the frequent and recurrent toss and catch failures and the physical needs to chase and collect objects. Such events reduce the effective time of learning during a training session and impair trainee motivation. Taken together, insufficient processing and memorization, inefficient use of training time, and reduced motivation all decrease the efficiency of juggling training in the real world. One possibility enabled by the LWJ is a general slowing or speeding up of the overall time clock (pace) of the task (similar to a slow or fast motion of a video clip). Such slowing allows trainees to observe, perform, and practice the task in "slow motion." When speed is increased the pace of performing the trick may now be even faster than normal. Note that this option that exists in the virtual but not in the real world does not change any of the relationships between elements, the pattern, or the order in which they have to be responded to. However, trainees get the necessary time to recognize patterns, execute actions, and commit them to memory. In contrast, when speeding up performers are required to cope with task demands that are faster than normal.

In the present study (Erev, Gopher, Lammfrom, & Lippi, in preparation) students with no prior juggling experience trained in the LWJ on three-ball cascade juggling for eight sessions, each made up of ten 4-minute practice trials, under task speed change. There were six different speeds: one was the normal, three were slower, and two were faster than normal. Practice started at slow speed and trainees graduated to faster speed levels when meeting the required graduation criterion of two consecutive juggling cycles within a 4-minute practice trial. Training starting from slow speed was expected to allow subjects to develop a better representation of the trick pattern. Juggling performance of the speed change group was compared on a ninth test session performed at normal speed to a control group that practiced for eight sessions only at normal speed. Performance was measured by the average representing the longest sequence of balls caught and tossed without a fall within a 4-minute block. An additional 10th session was given when both experimental and control group were required to perform two new tricks: reversed cascade and snake. The results show that the group trained with speed changes obtained significantly higher juggling scores than the control group on the ninth test session (14.86 versus 8.75, respectively). In addition, subjects trained under speed change obtained significantly higher scores when transferred to the reversed cascade trick, which is a variant of the original trick they were trained on (7.00 versus 3.67, respectively). No differences were observed on the transfer to the snake trick which is very different and much more difficult.

EMPHASIS CHANGE IN TRAINING FOR CONTROL OF THE SPATIAL-TEMPORAL BALANCE OF TOSS AND CATCH OF BALLS

Development of rhythmic stable juggling, by mastering and controlling the required tossing and catching balls spatial-temporal relations, is another major requirement that novices are faced with while learning to juggle. A stable spatial-temporal ratio is a direct derivative of ball time in the air (toss height) and time in hand. A second study in the LWJ was directed to teach trainees control strategies to enable them to change and cope with different spatial-temporal balance points. The manipulation used in this experiment is the emphasis change methodology (Gopher, 1994). Emphasis change is a training protocol in which trainees are required during training to systematically change their voluntary emphasis (effort, attention allocation) on subelements of a performed task. Emphasis levels are varied between few minute practice trials or performance durations (Gopher, 2007). Experiments and application studies have demonstrated the robustness of this protocol in improving the ability of performers to cope with high workload tasks. In the present study trainees were asked to voluntarily change their emphasis either on height of the three balls juggling pattern, or the time duration that each ball is held in hand between catch and toss. Three groups of students with no prior juggling experience practiced the three-ball cascade juggling for 10 sessions under different emphasis change manipulations. A fourth group had 10 sessions of regular practice without any emphasis instructions. Of the three experimental groups, one practiced with five levels of required ball height (H) (different required toss height levels indicated by a bar on the screen). A second group practiced with five different durations of holding balls in hand before tossing (D), as indicated by the tone pace of a metronome. The third experimental group received both manipulations first, separately, and then two sessions were performed with between 4-minute blocks change of emphasis on height or ball holding durations of ($H + D$). Juggling performance levels of the four groups were compared on the ninth and tenth sessions, both of which performed without any emphasis change under standard speed and height conditions. As in the previous speed manipulation experiment, sessions 11 and 12 were added, in which all groups were transferred to the new tricks of reversed cascade and snake.

All emphasis change group experimental groups obtained better juggling scores than the control group. The combined manipulation $H + D$ group, who practiced under both height and duration emphasis manipulations, was best at test (days 9 and 10). It hence appears that the emphasis change training method had a powerful contribution to the acquisition of the virtual juggling skill. It appears that training to juggle where the critical elements of "ball time in air" (height) and "ball time in hand" (duration) are emphasized, allowed subjects to develop better competence and control of juggling. Most pronounced was the improvement of trainees who were given both manipulations in combination.

TRANSFER OF TRAINING FROM THE LWJ TO REAL-WORLD JUGGLING

Because the LWJ lacks the motor and haptic aspects of juggling with real balls, the goal of the third study was to evaluate the relevance and value of the skill components that can be trained in the LWJ, for the acquisition of real juggling. The main question was whether training the pattern memory and the temporal-spatial relations of the three-ball cascade, which was shown to improve control and flexibility in virtual juggling performance in the LWJ, benefits the acquisition of real juggling skills. Participants practiced the acquisition of the three-ball cascade juggling (3BC) trick for 10 days. Real balls juggling performance was measured by the highest number of consecutive successful juggling cycles during a 4-minute practice trial. Participants in the control group practiced 3BC juggling during 10 sessions each containing eight 4-minute practice trials. The experimental groups had only five sessions of training with real balls, and the other five were conducted in the LWJ. Juggling under speed increase was administered on sessions 2 and 3, and emphasis change manipulations emphasizing either height or duration on session 5, and 7. Session 9 was conducted with a mix of high and duration changes. Subjects in the experimental group practiced with real balls only during sessions 1, 4, 6, 8, and

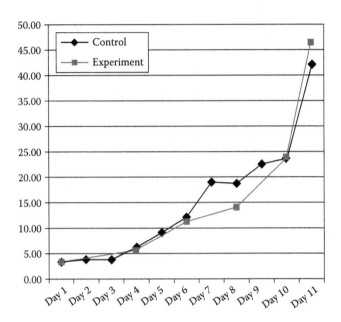

FIGURE 12.3 Mean 4-minute highest number of consecutive juggling cycles with real balls.

10 (the test session). On day 11, both groups were tested on juggling with real balls while instructed to perform tossing high or catching duration changes in different specific 4 min. juggling trials.

Results show that the experimental group trainees who shared their time between real ball training and the LWJ obtained comparable juggling performance levels to the control group trainees who were trained for all 10 sessions with real balls. Figure 12.3 depicts the two groups' mean highest number of consecutive cycles of juggling real balls, over the course of 10 days. During the 11th session, trainees who were previously given LWJ training were better able to follow the instructions to change their juggling performance to cope with required toss and catch changes. Figure 12.4 shows the mean highest number of consecutive cycles under normal conditions and when changes in catch duration and toss height were called upon.

DISCUSSION AND CONCLUSIONS

The first study evaluated the efficiency of the LWJ augmented by two types of audio-tactile stimulations. We examined whether audio-tactile events could help to acquire juggling skills that would transfer to juggling with real balls. We found that audio-tactile periodic stimuli in the LWJ did not improve significantly the performances in juggling with real balls when compared to the control training condition. However, the results suggest that the closed-loop implementation outperformed the open-loop solution. This result requires further studies to be validated but points at an interesting direction—that is, the control of the multimodal augmented reality in a light weight training platform should be directly based on the current behavior of the

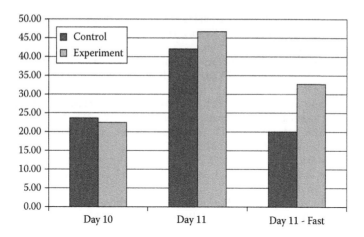

FIGURE 12.4   Mean highest number of consecutive juggling cycles on day 11, under normal conditions, and under required catch duration and toss height changes.

trainee, in a closed-loop scheme, and its timing properties should not be planned a priori. In three evaluation studies we examined the role of the cognitive components of the juggling skill when directly addressed in a light weight training platform transfer of training to juggling with real balls. The results obtained support the value of the LWJ as a virtual reality training platform and show that subjects who received only half of their training time with real balls were not inferior in their juggling performance and were better able to apply voluntary changes in juggling format as compared with subjects trained twice the time with real balls. These results now encourage the use of the existing LWJ platform to further understand the values and limits of simplifying the training of complex skills, and how such a training platform can be improved and modified to be used in an education or rehabilitation setting.

REFERENCES

Anderson, J. (Ed.). 1981. *Cognitive Skills and Their Acquisitions.* Mahweh, NJ: Erlbaum.

Anderson, J.R. 1992. Automaticity and the ACT theory. *American Journal of Psychology*, 105(2): 165–180.

Beek, P.J. 1992. Inadequacies of proportional duration model: Perspectives from dynamical analysis of juggling. *Human Movement Science* 11: 227–237.

Beek, P.J., and Lewbel, A. 1995. The science of juggling. *Scientific American* 273: 93–97.

Beek, P.J., and Turvey, M.T. 1992. Temporal patterning in cascade juggling. *Journal of Experimental Psychology: Human Perception and Performance* 18: 934–947.

Beek, P.J., and van Santvoord, A.A.M. 1992. Learning the cascade juggle: A dynamical systems analysis. *Journal of Motor Behavior* 24: 85–94.

DeGuzman, G.C., and Kelso, J.A.S. 1991. Multifrequency behavioral patterns and the phase attractive circle map. *Biological Cybernetics* 64: 485–495.

Draganski, B., Gaser, C., Busch, V., Schuierer, G., Bogdahn, U., et al. 2004. Neuroplasticity: Changes in grey matter induced by training. *Nature* 427: 311–312.

Erev, V., Gopher, D., Lammfrom, R., and Lippi, V., in preparation.

Gopher, D. 2007. Emphasis change as a training protocol for high demands tasks. In *Attention: From Theory to Practice*, Ed. A. Kramer, D. Wiegman, and A. Kirlik. Oxford: Oxford Psychology Press.

Gopher, D. 1994. Analysis and measurement of mental workload. In *International Perspectives on Cognitive Sciences*, Ed. G. d'Ydewalle, P. Eelen, and P. Bertelson, 265–291. London: Lawrence Erlbaum.

Haibach, P.S., Daniels, G.L., and Newell, K.M. 2004. Coordination changes in the early stages of learning to cascade juggle. *Human Movement Science* 23: 185–206.

Huys, R., Daffertshofer, A., and Beek, P.J. 2004. Multiple time scales and multiform dynamics in learning to juggle. *Motor Control* 8: 188–212.

Kelso, J.A.S., and DeGuzman, G.C. 1988. Order in time: How cooperation between the hands informs the design of the brain. In *Neural and Synergetic Computers*, ed. H. Haken, 180–196. Berlin: Springer.

Lagarde, J., and Kelso, J.A.S. 2006. The binding of movement, sound and touch: Multimodal coordination dynamics. *Experimental Brain Research* 173: 673–688.

Logan, G.D. 1988. Toward an instance theory of automatization. *Psychological Review* 95: 492–527.

Peper, C.E., Beek, P.J., and Van Wieringen, P.C.W. 1995. Coupling strength in tapping a 2:3 polyrhythm. *Human Movement Science* 14: 217–245.

Post, A.A., Daffertshofer, A., and Beek, P.J. 2000. Principal components in three-ball cascade juggling. *Biological Cybernetics* 82: 143–152.

Tuller, B., and Kelso, J.A.S. 1989. Environmentally specified patterns of movement coordination in normal and split-brain subjects. *Experimental Brain Research* 75: 306–316.

Zanone, P.G., and Kelso, J.A.S. 1992. Evolution of behavioral attractors with learning: Nonequilibrium phase transitions. *Journal of Experimental Psychology: Human Perception and Performance* 18: 403–421.

Zelic, G., Mottet, D., and Lagarde, J. 2012. Behavioral impact of unisensory and multisensory audio-tactile events: Pros and cons for interlimb coordination in juggling. *Plos One* 7(2) e32308. doi:10.1371/journal.pone.0032308.

13 Designing a Virtual Reality Training Platform for Surgeons

Theoretical Framework, Technological Solutions, and Results

Sylvain Bouchigny, Christine Mégard, Florian Gosselin, Pablo F. Hoffmann, and Maria Korman

CONTENTS

INTRODUCTION

Many socioeconomical factors influence the way in which surgical education is given to junior residents. Public consideration for patient safety has raised attention to the risks of residents practicing their skills on patients, and recent policies on workload limitations have increased the challenge of healthcare education by reducing the time available for teaching. To address these problems, new structures dedicated to postgraduate residency education which take advantage of technologies in surgical simulation* are appearing at a fast pace. New protocols are being developed to build

---

* Examples among many others are the SSC (Surgical Skills Center) at University of Southern California and the IRCAD (Research Institute against Digestive Cancer) in France.

the theoretical framework for the assessment of basic skills in surgery in which the role of surgical simulation can be addressed (Seymour et al., 2010).

Surgical simulation brings many challenges in education science and in technological research. The benefits of this approach gain credibility every year as more and more systems are being developed and tested. Of course, assessing the validity of the simulation in terms of transferring quality to reality is still a challenge and a key aspect of the present research. However, simulated systems, and in particular, virtual reality (VR) simulators, have critical features that already bring great pedagogical benefits. Simulation allows better control over pedagogical programs with higher repetition rates of a given task prior to the first practice in operating room. During the training, every parameter can be recorded and analyzed in real time leading to a much richer insight in terms of performance evaluation. A virtual environment (VE) is hence a tool for the selective combination of relevant or desired configuration that may affect perception and performance in the long run. Recent fundamental research suggests that proper selection and fine tuning of the relevant sensory modalities involved in procedural learning and perception in VE can facilitate acquisition of basic skills in surgeries involving strong haptic feedback (Tsuda et al., 2009).

Within the SKILLS project unified framework, we have developed a surgical training platform following a user-centered approach along with strong technological developments. The goal was to evaluate the benefits of features commonly available in VR surgical simulation to the training of skills in maxillofacial surgery (MFS). We focused on a surgery in which haptic basic skills are of primary importance, the Epker osteotomy. This surgery requires highly controlled drilling with a major reliance on haptics over visual interaction. Within this context, a training protocol based on modality management during jaw drilling procedures was developed and tested in university hospitals. The novelty of our work is in the design principles of the training platform, where the priority was given to the way sensory feedbacks are combined when presented to the subject. Such a dissociation and control of sensory-motor features is impossible in a real environment, where sound, vibration, and stiffness naturally come together as a complex input, originating from the same physical source during the bone drilling procedure. Another important feature implemented and tested is the possibility to record selected parameters during the procedure and process them to provide both the trainee and the trainer with a detailed and quantitative scoring of the performance. Together, these features along with significant technical developments signify the uniqueness of the MFS platform that allows for the transfer of procedural and sensory expertise from expert surgeon to a trainee in a highly controlled, measurable, and individualized manner.

The MFS training platform focuses on mastering drilling procedures. This ability is rarely included in traditional curriculums (at least not in the very particular context of mandibular corticotomy where critical anatomic parts must be avoided), and residents usually practice and learn these skills in actual surgeries. It is a task for which a VR training system can be of great benefit. In the following sections we describe the critical steps in the development, design, and evaluation of the MFS platform and the training protocol. We also summarize the main results of our studies, which are published separately (Bouchigny et al., 2011; Korman et al., 2011).

FIELD STUDIES AND ERGONOMICS

Epker osteotomy is a maxillofacial surgery performed to correct tooth alignment between the upper and lower maxilla in the case of malformation, disease, or trauma. The principle of this surgery is to drill the cortical bone of the lower maxilla on both sides along four lines on each side located on the hardest part of the skull (see Figure 13.1c). The maxilla is then fractured (i.e., the main part of the mandible body is separated from the ramus on both sides by a distraction movement). Finally, the maxilla is fixed in a corrected position. Throughout the entire procedure, the integrity of the trigeminal nerve running through the mandible and responsible for the labial sensitivity must be guaranteed. This surgery is considered by the faculty as representative of surgical procedures difficult to master and teach with traditional techniques. MFS does not yet benefit from the development of educational simulators such as in dentistry (Cormier et al., 2010; Kim and Park, 2009; Tse et al., 2010) and minimally invasive surgery (Richards et al., 2000). Also, masteoectomy training simulators have already been developed in the laboratory for temporal bone surgery (Morris et al., 2006) with the objectives to help young surgeons exercise their procedural knowledge. However, education in MFS still refers to William Halsted's model (Osborne, 2007) of surgical education with observation and repetition from observed experts.

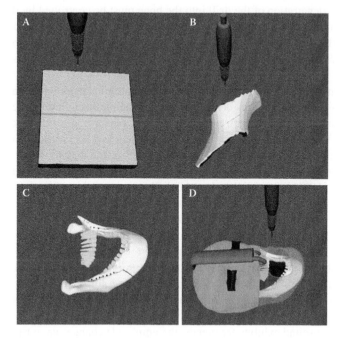

FIGURE 13.1 *(See color insert.)* Example of exercises: (a) drilling a line on a flat surface (SK protocol); (b) drilling a line on a curved surface (SK); (c) line of the Epker osteotomy seen on a maxilla (SI); (d) drilling the cortical with tissue interaction (last exercise for both SK and SI).

Building a skills trainer requires the identification of surgical expertise as well as mandatory skills that must be mastered by residents before entering the operating room. We developed a methodology aimed at the identification of skills required in MFS. A training protocol of the main skills involved in this surgery was defined utilizing a user-centered approach. The design process relied on field studies, interviews, and workshops with surgeons. Surgeons from different institutions participated in the interview to cover possible variable techniques, with special emphasis on the critical steps of the task. Workshops with surgeons and the technical team were organized regularly during the project to elaborate and discuss training protocol of those specific skills. The interviews were then complemented with a field analysis in theater. A surgery that was performed by an expert senior surgeon was fully analyzed. The results of this analysis were then formalized into a task model that was discussed and validated during a meeting with the surgeons and knowledge of the skills used to perform the task following (Ley et al., 2010). The most difficult subtasks were found to be related to bone interaction, requiring very delicate haptic abilities that can only be trained with focused and repeated practice. After the incision of the inner cheek and the removal of the periosteum, the surgery starts with the location of the Spix's spine to protect the entrance of the trigeminal nerve from the first drilling line. The location of the spine relies heavily on haptic shape perception when visual feedback is restricted by the anatomy of the skull. Then the nerve is located and protected with a tool and the surgeon can proceed to the corticotomy. The penetration of the drill into the bone must be controlled by the surgeon in a much more precise way than what is usually the case in orthopedic surgeries. Controlled drilling in space limited to the cortical part of the bone requires application of forces up to 30 N and strict compliance management while the visual field is occluded or limited.

Complementary measures were performed in anatomic laboratory to gather quantitative data that were used as design drivers for the platform developments. Three expert surgeons, equipped with tracking systems, performed a full Epker surgery on three different cadaver heads. Apart from getting key specifications needed for the technical development of the platform, analyses of the data provided some very interesting insights of the surgery. We analyzed the variation of sound and force during punctual drilling, especially when the drill is entering the inner part of the bone going from a high-density to a low-density medium. This is a sensation the surgeon must anticipate and detect to avoid penetrating too far into the spongy bone. One of the most important parameters in our study is stiffness, which measures the variation of force as the drill is moving forward. The force profile typically has shown to have three phases: stiffness is kept constant by the surgeon while drilling the cortical bone (meaning that the surgeon is increasing its force linearly as a function of the distance through the bone), an inflection is seen when the burr starts to leave the cortical, and, finally, a rapid drop is observed when the burr reaches the spongy bone. In parallel, the sound exhibits frequency and amplitude changes upon rupture of the cortical. During drilling of the cortical, vibration and sound are mainly composed of the fundamental frequency corresponding to the rotation speed of the drill and its first harmonics. But when the burr leaves the cortical bone, the first harmonics completely disappears. Surgeons are very likely to use these complex modulations in sensory feedback upon application of force to control progress, stop, and withdrawal of the drilling tool (see, for example, Praamsma et al., 2008, for the importance of sound). For that reason the platform was designed to render these parameters with high resolution. Given the importance of fast sensory modulations in the critical steps of the MFS surgery, psychophysical experiments were conducted to find possible perceptual accelerators that promote better acquisition and transfer of relevant perceptual skills (i.e., stiffness and vibro-auditory discrimination). To this end the role of multimodal stimuli presentation and definition of optimal order of sensory experience for different training conditions in VE were addressed in a series of systematic studies, presented in the next section.

TOUCH SCIENCE: FUNDAMENTAL RESEARCH IN MULTIMODALITY MANAGEMENT

How multimodal inputs unite to form an integrated sense of an object is still an open issue in neuropsychology. During the examination of an object, the haptic system provides information about tactile information as well as information about arm displacement in conjunction with signals of applied force (i.e., kinesthesis information) (Clark and Horch, 1986). Yet information regarding the displacement or deformation of an object may also be obtained from the visual system, which adds valuable information over time if afforded (Lederman and Klatzky, 2009). In contrast, auditory-tactile perception of vibratory stimulation is a unique type of multisensory interaction in which the two sensory modalities involved are sensitive to the same kind of physical property—mechanical pressure in the form of oscillations, where the very same vibratory stimulus can be experienced simultaneously by the peripheral receptor organs of both sensory modalities (the basilar membrane in the cochlea and the skin) (Soto-Faraco and Deco, 2009). The human somatosensory system thus integrates tactile and kinaesthetic signals along with visual and auditory information. The degree of sensory dominance in multisensory events is suggested to be determined by the statistical reliability of the available sensory information (Ernst and Banks, 2002). Little is known about optimal conditions for long-term training of touch aspects of haptic tasks (Gallace and Spence, 2009). Auditory-visual studies argued for a facilitatory effect of multisensory training on unisensory learning (Seitz et al., 2006). As well, information feedback is known to have an enhancing effect on performance (Adams, 1987). These ideas needed to be systematically addressed in touch perceptual training.

Tool-tissue interactions during drilling are of complex multisensory nature: changes in bone structure should be rapidly evaluated to detect the transition from stiffer, cortical bone layer to inner spongy layer. During the course of the drilling procedure (that is very fast, around 1 or 2 seconds), the visual field may be occluded by surrounding tissues, such that correct surgical performance is primarily based on haptic sensory information. The surgeon, regardless of the properties to which he or she intentionally attends, may use any or all types of information available (stiffness, vibrotactile, and accompanying auditory sensory modulations during drilling) to detect and discriminate the bone depth at which drilling should be stopped. Mistakes will lead to either superficial or incomplete drilling or to excessively deep drilling that may harm the nerve inside the spongy bone.

Motor-cognitive and perceptual capabilities that are essential for the MFS procedure were evaluated in the context of the nature of the haptic experience, matching of modalities, and order of sensory combinations within and between training sessions. The following critical questions that pertain to multisensory stimuli perception and memory were addressed in several empirical studies run on the MFS platform or using simpler off-the-shelf VE setups:

> What are the within-session effects of order of experience and sensory modality in stiffness discrimination?
> What is the time-course of long-term (between days) learning in unimodal stiffness discrimination?
> How do multimodal training and information feedback affect unimodal learning?

The experiments were conducted using a touch-enabled computer interface capable of providing users with visual and haptic stimuli differing in the stiffness intensities presented (PHANToM Desktop), using a two-alternative forced-choice (2AFC) stiffness discrimination task. Subjects discriminated between the stiffness of two targets that were presented either haptically or visual-haptically in two subsequent blocks, and their performance was measured in terms of the proportion of correctly discriminated pairs of targets. Our results showed that prior training in a unisensory haptic stiffness discrimination task greatly improved performance when visual feedback was subsequently provided. This improvement in performance extended beyond the practice effect seen when the unisensory haptic task was repeated twice, or when the multisensory task was repeated twice, or when the multisensory task was followed by the unisensory haptic task. Moreover, this effect was only evident for supra-threshold stimuli pairs. Our findings suggest that optimization integration theories of multisensory perception need to account for past sensory experience that may affect current perception of the task (Korman et al., 2011).

In the second stiffness discrimination study participants were trained in two consecutive days and were tested on haptic-only condition before and after training. During training two variables were manipulated between participants: knowledge of results (KR) and addition of congruent visual information. Results showed that the time-course of unisensory stiffness discrimination learning was specific to the combination of sensory information during training, and in bimodal training conditions did not generalize fully to post-training unisensory performance. In addition, information feedback during training positively affected decision time only during training and has not been shown to be effective in improving the accuracy of discrimination.

Thus, we have shown that the amount and type of sensory information that are traditionally in the focus of multi modal integration studies, are not the only critical factors affecting stiffness discrimination ability, as even a within-session manipulation of order of experience has a robust effect on the way that visual and haptic sensory cues fuse. In line with recent notions in cognitive and neural correlates of tactile memory (Gallace and Spence, 2009), our findings suggest that early presentation of unisensory tactile information may help to retrieve the multisensory qualities of the objects that they interact with haptically. This result was taken as an accelerator in the definition of training protocol, to account for the inherited uncertainty about the amount of visual information afforded during jaw drilling. In the evaluation studies this approach has been proved to be beneficial in terms of training drilling subtasks on the MFS platform.

How does the nature of stimulus presentation—uni- or bimodal—affect vibro-auditory amplitude perception?

What are the similarities and the differences in the perceptual ability of expert surgeons compared to young naive participants?

In this study, run on the MFS platform, we explored the ability to detect and discriminate changes in vibrotactile stimuli amplitude based either on purely haptic feedback or together with congruent synthesized auditory cues in naive participants and expert surgeons.

We showed that in discrimination tasks, bimodal performance was always better than uni-modal performance regardless of order of experience. The discrepancies between the results on vibration and stiffness studies may reflect the inherited physiological differences in perception of these two aspects of touch: stiffness is commonly sensed without vision; however, vibration of a tool is always accompanied by sound, where the very same vibratory stimulus simultaneously activates receptors on the basilar membrane in the cochlea and the skin of the fingers. Experiments with surgeons revealed that expertise in complex skill of MFS strongly relies on enhanced bimodal touch perception, as measured in reaction times and discrimination ability in vibro-auditory conditions. These results suggest that acquisition of mandibular surgery skill has brought an enhanced representation of vibro-tactile modulations in relevant stimuli ranges.

PLATFORM IMPLEMENTATION

The studies presented above were translated into a training system specifically designed to reach the needs of a maxillofacial surgery in terms of workspace, force, torque, and stiffness while implementing training accelerators with the benefits of new possibilities brought by virtual environment. The maxillofacial surgery SKILLS training platform implements high-performance components necessary for the multimodal rendering of the visual, audio, and haptic information involved in the simulated procedures and associated training program. All elements are integrated in a mobile cabinet that can be moved easily to hospitals and training places. The upper part of the platform incorporates the input/output devices manipulated by the user (see Figure 13.2), while the lower part is used to carry all computers and electronics running the VR simulations and devices' controllers. To ensure a seamless transition from the VR training environment to the real one and thus favor the transfer of the acquired skills from the simulation environment to the operating room, the platform aims at reproducing this information with a high level of realism. The platform makes use of a newly developed high-fidelity haptic device specifically optimized

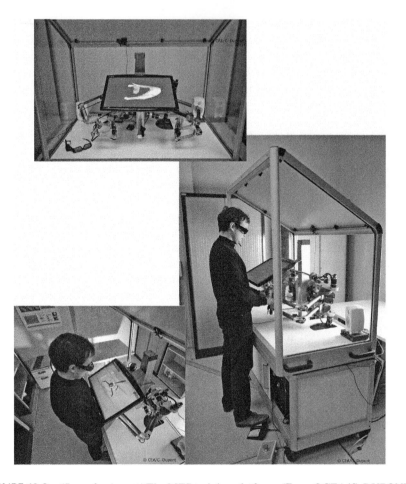

FIGURE 13.2 *(See color insert.)* The MFS training platform. (From ©CEA/C. DUPONT.)

to simulate and finely reproduce interaction forces between the surgery tools with bones and tissues of the virtual patient.

To ensure both a high-quality force feedback and dexterous manipulation within a large orientation workspace, a hybrid serial/parallel robotic architecture is used on this device. The useful workspace is larger than $162 \times 194 \times 177$ mm in translation and more than $140°$ around all axes in rotations as specified from real surgeries, while at the same time the maximum forces and control stiffness are larger than 50 N and 10,000 N/m along all axes (Gosselin et al., 2011). The device is further equipped with an active handle powered by a piezoelectric stack actuator allowing up to 100 μm of normal surface displacement at 600 Hz (Giuntini et al., 2010). Another device featuring a similar architecture previously designed for minimally invasive telesurgery is used to manipulate the vacuum cleaner with the left hand. Its workspace is similar but its force capacity and control stiffness are only 15 N and 5000 N/m. This is sufficient to simulate the interactions between the vacuum cleaner and the soft tissues of the mouth but not for drilling.

The platform integrates a 22-inch three-dimensional (3D) liquid crystal display (LCD) monitor. The user wears active LCD shutter glasses and has a 3D view of the operating site. Moreover, the glasses are equipped with a motion capture tracker whose basis is attached to the screen. This way, the movements of the user's eyes are measured, and the visual rendering is adjusted to remain fixed regardless of the user's head movements. Finally, the training platform makes use of active loudspeakers allowing high-fidelity audio feedback. The sound and vibration of drilling was synthesized using a sinusoidal model (Hoffmann et al., 2009, 2009a). Sinusoidal parameters corresponding to frequency, amplitude, and phase were extracted from audio recordings using a matching pursuit algorithm. The minimum number of sinusoids required for a good perceptual approximation for audio feedback was found by using a model of simultaneous auditory energy masking (van der Heijden and Kohlrausch, 1994). Based on the output of the model in combination with the results from a psychoacoustic experiment, we found that on average 30 sinusoids were needed for a perceptually adequate synthesis of the drilling sound (see Chapter 7 for further details).

A representative sample of drilling was selected and synchronized with the corresponding force recording in order to establish a mapping between force and the sinusoidal parameters. The general criterion to select the force signal was to maximize the range spanned by the force values starting from the lowest possible value and increasing as monotonically as possible to the largest possible value. Because no clear systematic relation was found between force and sound parameters, a look-up table was constructed with frequencies and amplitudes arranged as a function of force using a grid spanning the range from 0 to about 10 N. The resolution of the grid was not constant having a higher resolution for smaller force values. The force signal used to map force values with sinusoidal parameters was represented by a third-order polynomial because we found this fit to be in agreement with the denoised original force signal (R-squared = 0.99). Sinusoidal phase information was randomly generated from a uniform distribution between $[-\pi/8, \pi/8]$. This process for generating phase information added the typical noisy characteristic of drilling sounds while still keeping their tonal structure.

All training exercises were implemented in the XDE framework developed at CEA-LIST for real-time simulation. Both applications run on a high-performance workstation (ref. HP XW8600, bi-quad core Intel Xeon 5450 running at 3 GHz, Quadro FX5600 graphic card). This computer is connected to an external sound card controlling the loudspeakers (ref. Edirol FA-66, 24 bits, 192 kHz), to a power amplifier connected to the piezoactuator (ref. Piezomechanik LE150/100/EBW) and, via ethernet, to the haptic interfaces controllers (ref. Haption 9 axes, 1600Hz).

TRAINING PROTOCOLS AND EVALUATION

The MFS platform training program focuses on mastering drilling procedures. The exercises consist of either performing a punctual drilling or drilling a line which are two of the basic skills needed to do the corticotomy. The overall protocol is designed so that it does not require any deep anatomical knowledge and procedural competences. In this way, the protocol can be followed by both junior and senior residents.

The sensory feedback involved in such exercises is composed of three main dynamic components: a haptic component (compliance and vibration) because the sensation varies abruptly from hard to softer material, a visual component as the movement of the drill can be perceived visually, and a sound and vibration component as the drilling sound varies according to the material and the position of the drill.

In a virtual environment, different approaches might be considered to design this type of exercise: simulate the whole procedure and allocate trainees some time on the platform to train their skills as if they were in the operating room, or simulate only key elements at each step of the training protocol allowing the trainee to concentrate on specific skills. With this approach, the overall simulation does not need to have the same level of detail but will focus on subskills, thus concentrating more on the key aspects of perceptual and motor control learning. On the MFS platform, the two points of views were considered, two different protocols were developed, and their efficiencies were compared.

The two protocols were segmented into nine exercises performed in three training sessions of approximately 45 to 55 minutes spanning over 20 days. During the course of the training, each trainee will have drilled at least 300 holes and 20 lines. The first type of training, based on the simulation protocol (or SI), consists in practicing basic skills directly on a physiologically realistic jaw, performing progressively the different steps of the surgery. The second approach, called the skills protocol (or SK), exploits VR environment features beyond simple simulation to help the trainee to focus on the identified key procedural and perceptual skills with progressively increasing complexity and combination of features. Figure 13.1 gives some examples of the different shapes implemented in the exercises from simple to more complex and realistic. Following the results in fundamental research presented above, a multimodal management approach is implemented in the first three exercises where the abilities of punctual drilling on a flat surface are learned. These exercises were provided to the trainee in the following order:

- Step 1: Haptic only
- Step 2: Haptic + sound + vibration
- Step 3: Full set of modalities is provided together: visual, audio, tactile, and force feedback

Three performance criteria were retained for this study: (1) distance penetration below the cortical in the spongy bone, as it is a key parameter to master in order to protect as much as possible the trigeminal nerve inside the jaw; (2) time to complete the drilling (mean and standard deviation) as a way to evaluate the efficiency of the task; and finally, (3) self-confidence level.

The protocols SK and SI have been studied during two experimental campaigns in the maxillofacial surgery department of CHU Amiens, France, and Hopital Roger Salengro in Lille, France. These campaigns had two purposes: validating the simulation quality by gathering feedback from surgeons and comparing the efficiency of the two protocols in a population of residents. For this purpose, a pre-test and a post-test of punctual drilling in simulation were added to evaluate the two groups on one common measurement. Overall, 20 residents followed the protocol, 10 in each group. These two groups included, each, six junior residents who never practiced osteotomy, and four senior residents who already practiced in the operating room. In addition, performances of four maxillofacial experts were involved to gather expert performance values.

The first result of interest is that expert surgeons achieved during the pre-test the highest score without any prior training on the platform. Their performances were comparable with the values recorded in the anatomic laboratory in the task analysis campaign. The result showed that junior and senior residents from both protocols performed very differently. Junior residents were fast but very inaccurate, while senior residents were more careful and they favored accuracy (depth of penetration) versus speed (time to complete). This fact is in line with the information extracted from the interviews—junior residents had no specific prior knowledge of what are the most important factors to be considered in the operating room. On the contrary, senior residents were already instructed during their curriculum that minimizing trauma in the inner bone is much more important than speed. Therefore they took the necessary time whenever required. Training induced significant improvements in both groups. Junior residents improved their speed and accuracy, while senior residents improved their speed but maintained the accuracy constant.

There was a significant interaction of the post-training gains regarding the type of training protocol. The improvement of speed was significantly better for the SK group. Also, the variance of stiffness* measured for a set of drills during pre- and post-test decreased significantly for the SK group and became comparable to the expert group, while it stayed constant for the SI group leading to the conclusion that SK group became more constant in their strategy to perform the drill. Finally, self-confidence was raised from before to after the training only for junior residents. Senior residents kept similar estimates between pre- and post-self-confidence measurements.

SUMMARY AND FUTURE DIRECTION

Our methodology provided qualitative and quantitative description of major sub-skills constituting the MFS surgery. Observational methodologies and analysis from human science successfully obtained information on surgeons' routine procedures and helped them to elaborate meta-knowledge of their procedures and skills. The user-centered approach and participative design followed during the course of the project showed its efficiency as part of the design methodology for elaborating a skills trainer in MFS. Results have shown in both fundamental research experiments and field studies that unimodal stimulation prior to multimodal stimulation brings benefits in basic skills training programs. In addition the platform has been success-fully tested to reflect the expertise of a subject, as expertise in the operating room is directly correlated with performances on the platform's training protocol. Future work should include a more detailed evaluation of the transfer to the real operating room of the skills acquired on the platform.

---

* See section "Field Studies and Ergonomics" for a definition of stiffness in this case.

ACKNOWLEDGMENTS

The authors warmly thank P. Delcampe, C. D'Hauthuille, and F. Taha for their cooperation in the design process of the system, and they express their gratitude to B. Devauchelle and J. Ferri for their valuable help and support to the experiments.

REFERENCES

Adams, J.A. (1987). Historical review and appraisal of research on the learning, retention, and transfer of human motor skills. *Psychological Bulletin*, 101, 41–74.

Bouchigny, S., Mégard, C., Gabet, L., Hoffmann, P., and Korman, M. (2011). Evaluation of a multimodal VR training platform for maxillofacial surgery. In *BIO Web of Conferences*, Volume 1, The International Conference SKILLS 2011, Montpellier, France.

Clark, F.J., and Horch, K.W. (1986). Kinesthesia. In K. Boff, L. Kaufman, and J. Thomas (Eds.), *Handbook of Perception and Human Performance. Cognitive Processes and Performance* (1986). Volume: 1, Issue: 5, Publisher: Wiley, pp. 13-1 to 13-62.

Cormier, D., Pasco, C., Syllebranque, C., deKeukelaere, P., and Chavallier. (2010). VirTeasy a haptic simulator for implantology surgical training designed by an activity analysis. *Proceedings of the 10th Virtual Reality International Conference* (VRIC), Laval (France).

Ernst, M.O., and Banks, M.S. (2002). Humans integrate visual and haptic information in a statistically optimal fashion. *Nature* 415: 429–433.

Gallace, A., and Spence, C. (2009). The cognitive and neural correlates of tactile memory. *Psychol Bull.* 135(3): 380–406.

Giuntini, G., Ferlay, F., Bouchigny, S., Gosselin, F., Frisoli, A., and Beghini, M. (2010). Design of a new vibrating handle for a bone surgery multimodal training platform. In *Proc. 12th Int. Conf. on New Actuators*, Bremen, Germany, pp. 85–88.

Gosselin, F., Ferlay, F., Bouchigny, S., Mégard, C., and Taha, F. (2011). Specification and design of a new haptic interface for maxillo facial surgery. In *Proc. IEEE Int. Conf. Robotics and Automation*, Shangai, China, pp. 737–744.

Hoffmann, Pablo F., Gosselin, F., and Taha, F. (2009). Analysis of the drilling sound component from expert performance in a maxillo-facial surgery. Paper presented at Proc. of the 15th Int. Conf. on Auditory Displays (ICAD), May 18–22, Copenhagen, Denmark.

Hoffmann, Pablo F., Gosselin, F., Taha, F., Bouchigny, S., and Hammershøi, D. 2009a. Analysis of the drilling sound in maxillo-facial surgery. In *Proc. of the International Conference on Multimodal Interfaces for Skills Transfer*, Bilbao, Spain, 15–16 December.

Kim, K., and Park, J. (2009). Virtual bone drilling for dental implant surgery training. In *Proceedings of the 16th ACM Symposium on Virtual Reality Software and Technology*, pp. 91–94 (Kyoto, Japan, November 18–20, 2009). S.N. Spencer, Ed. VRST '09. ACM, New York.

Korman, M., Weiss, K., Cohen, A., Reiner, M., and Gopher, D. (2011). Effects of practice and sensory modality on stiffness perception. *Presence.* Accepted.

Lederman, S.J., and Klatzky, R.L. (2009). Haptic perception: A tutorial. *Attention, Perception, and Psychophysics* 71(7): 1439–1459.

Ley, T., Kump, B., and Albert, D. (2010). A methodology for eliciting, modelling, and evaluating expert knowledge for an adaptive work-integrated learning system. *Int. J. Human-Computer Studies* 68, 185, 208.

Morris, D., Sewell, C., Barbagli, F., Blevins, N., Girod, S., and Salisbury, K. (2006). Visuohaptic simulation of bone surgery for training and evaluation. *IEEE Transactions on Computer Graphics and Applications*, November, 48–57.

Osborne, M.P. (2007). William Halsted: His life and contributions to surgery, *Lancet Oncol.*, Mar, 8(3): 256–265.

Praamsma, M., Carnahan, H., Backstein, D., Veillette, C., Gonzalez, D., and Dubrowski, A. (2008). Drilling sounds are used by surgeons and intermediate residents, but not novice orthopedic trainees, to guide drilling motions. *Canadian Journal of Surgery* 51(6): 442.

Richards, C., Rosen, J., Hannaford, B., Pellegrini, C., and Sinanan, M. (2000). Skills evaluation in minimally invasive surgery using force/torque signatures. *Surgical Endoscopy* 14, 791–798.

Seitz, A.R., Kim, R., and Shams, L. (2006). Sound facilitates visual learning. *Curr. Biol.* 16, 1422–1427.

Seymour, N., and Scott, D. (2010). *Simulation and surgical competency. An issue of surgical clinics,* 1st Ed., Vol 90-3, Saunders.

Soto-Faraco, S., and Deco, G. (2009). Multisensory contributions to the perception of vibrotactile events. *Behav. Brain Res.* 196(2): 145–154. Epub 2008 September 30.

Tse, B., Harwin, W., Barrow, A., Quinn, B., San Diego, J., and Cox, M. (2010). Design and development of a haptic dental training system—hapTEL. *Proc. Eurohaptics,* part II, 101–108.

Tsuda, S., Scott, D., Doyle, J., and Ones, D.B. (2009). Surgical skills training and simulation. *Current Problem Surg.* 46(4): 271–370.

Van der Heijden, M., and Kohlrausch, A. (1994). Using an excitation-pattern model to predict auditory masking. *Hearing Res.* 80, 38–52.

14 Training Platforms for Upper Limb Rehabilitation

Antonio Frisoli, Denis Mottet, Isabelle Laffont, and Massimo Bergamasco

CONTENTS

CURRENT ISSUES IN UPPER LIMB REHABILITATION AFTER STROKE

Cerebrovascular disease (i.e., stroke), represents the first cause of acquired permanent disability in adults and the third leading cause of death in industrialized countries. Even after rehabilitation, about 80% of post-stroke patients still suffer upper limb disability, which impairs their daily living activities and often leads to major incapacities (Kwakkel et al., 2006). The mechanisms of functional recovery after stroke remain largely unknown, but animal models and correlated human studies demonstrate that functional recovery after stroke is obtained through the use-dependent reorganization of neural mechanisms, exploiting basic properties of neural plasticity (Hallett, 2001).

Current practices in standard care after stroke mainly rely on manual techniques provided by therapists. The basis of such techniques is task-oriented training, which combines cognitive and motor rehabilitation (Langhorne et al., 2011). Importantly, such rehabilitation is to be intense, to start very early, and to be individualized depending on patient needs (Horn et al., 2005). The consequence is that clinicians need efficient strategies for both the assessment of patient recovery and the delivery of a high dose of training.

Robot-assisted arm training has the potential to be of great help in both cases. Robot-assisted arm training uses robotic technologies to accelerate the recovery of upper limb function, with positive results on strength and motor function of patients (Lo et al., 2010). Although the lasting effects on the quality of the activities of daily living of patients remain to be confirmed (Mehrholz et al., 2009), robotic technology offers the opportunity of new therapeutic strategies based on recurrent evaluation of motor function to optimize the dose of the training and tailor the rehabilitation program to patients' needs (Han et al., 2008).

From the perspective of therapists, robotic technologies provide somewhat unique opportunities toward higher volume and more controlled rehabilitation. However, it might be even more important that robotic technologies allow to monitor "for free" the time course of patient recovery, to give performance feedback to the patient, and allow for online multisensory feedback that may lead to a better retention of regained functions (Cameirao et al., 2009). Moreover, movement kinematics may be used to assess movement recovery in a more precise way (Roby-Brami et al., 2003; Subramanian et al., 2010), going far beyond clinical tests that often do not clarify how the movement has been built, and what is due to motor recovery or compensation. For example, in stroke patients, goal-directed movements are characterized by a decrease in velocity, an increase in the number of submovements, and abnormal muscular activation patterns (Mccrea et al., 2002). As a consequence, measuring movement smoothness not only informs about patient's capability of executing a given movement, but also about movement quality (Amirabdollahian et al., 2002).

In this chapter we present the last results of the robotic-based training and assessment for upper limb rehabilitation performed with a robotic exoskeleton and an innovative bimanual training system in hemiplegic patients after stroke.

ROBOTIC UPPER LIMB TRAINING WITH THE L-EXOS AFTER STROKE

A class of rehabilitation robots that deserve a particular interest in upper limb rehabilitation is constituted by active exoskeletons. Exoskeletons are robots with kinematics isomorphic to the human arm that can be worn on the user's arm. They present several significant advantages compared to an end-effector-based system. They may track the full arm kinematics, so not only the hand position, but the whole posture of the arm, and consequently apply force assistance at the level of each joint (e.g., shoulder, elbow, and wrist).

Because an exoskeleton can act on each joint independently, it offers the unique opportunity to re-train interjoint synergies, by constraining the shoulder-elbow coordination, for example. The workspace covered by these systems is three dimensional and large, so the training can be conducted on fully spatial movements and along different directions.

The Light-Exoskeleton (L-Exos) (Figure 14.1a) is a right-arm rehabilitation robot used for stroke patients' rehabilitation. It is composed of five degrees of freedom

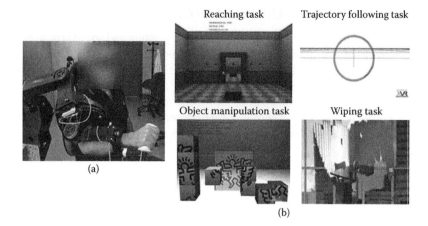

Reaching task   Trajectory following task

Object manipulation task   Wiping task

(a)

(b)

FIGURE 14.1 *(See color insert.)* (a) The Light-Exoskeleton Platform; (b) some training exercises proposed with the system.

(DoF), of which four are active (i.e., shoulder and elbow flexion/extension, shoulder internal/external rotation, and abduction/adduction), and can provide both passive and active assistance to the patient during the execution of movement.

The effects of a rehabilitation training conducted with the L-Exos within a group of hemiplegics after stroke was assessed in two clinical trials that made use, respectively, of a kinesiologic or robotic-based assessment.

EVALUATION OF EXOSKELETON-BASED TRAINING WITH A KINESIOLOGIC ASSESSMENT

METHODS AND PARTICIPANTS

Nine chronic stroke patients underwent treatment for 6 weeks, three times per week, with the L-Exos, consisting of a series of training exercises, shown in Figure 14.1b. In the simplest way the assistance provided by the robot consisted in a gravity counterbalancing of weight of the arm, as this has been proved to enlarge the workspace of the arm and to be effective for motor recovery. Alternatively, the robot provided guided assistance to the task to be performed, according to an impedance-based model.

At admission and discharge, patients were clinically evaluated with Fugl-Meyer assessment and Modified Ashworth scales, respectively, assessing global sensory-motor recovery and spasticity. Moreover, all patients underwent a kinesiologic evaluation performed during the execution of reaching movements toward targets placed in the horizontal plane, to assess both execution time and smoothness associated to movement execution.

Smoothness of movement was evaluated on the basis of the kinesiologic measurements, analyzing the velocity profile associated to the ulna displacement, with a smoothness index (SI) defined as the count of local minima in the velocity profile.

ASSESSMENT OF RECOVERY WITH THE L-EXOS SYSTEM

As a practical example, in Figure 14.2 we show the position and speed profile of the ulna movement of one patient during a reaching movement performed before (left panel) and after (right panel) the robotic training. It is clear that the velocity profile is indicative of the improved movement execution, and that the double peak of the velocity profile is clearly regained after the robotic training.

We experimentally found in chronic stroke patients that the smoothness index can predict both Fugl-Meyer clinical assessment and Modified Ashworth assessment, as shown in Figure 14.3.

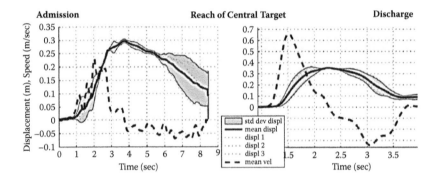

FIGURE 14.2 Typical time series of position (mean continuous line) and velocity (dashed line) during a reaching movement executed before and after robotic training. (From Frisoli et al. 2012.)

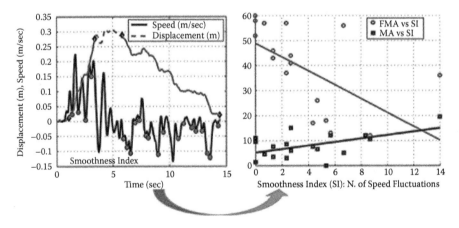

FIGURE 14.3 Correlation of smoothness index (SI) associated to movement against Fugl-Meyer and Modified Ashworth assessment scales.

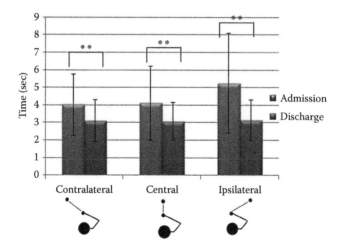

FIGURE 14.4   Time execution of movement before and after training for different movement directions (**$p < 0.01$).

RESULTS

We found significant improvements of Fugl Meyer scale (from 35.8 ± 18.2 before training to 40.3 ± 17.6 after training, $p = 0.006$) and overall reduction of the modified Ashworth scale (from 9.6 ± 4.5 to 6.3 ± 5, $p = 0.001$), indicating, respectively, an improved sensory-motor recovery and reduced spasticity. Moreover, also at functional level, we were able to notice better coordination in bimanual tasks execution (Frisoli et al., 2011), as reflected by assessment with the bimanual activity test (BAT).

It is moreover important to highlight how these changes were associated to better performance in movement execution. For instance, smoothness and execution time change in a different way for reaching movements performed along different directions of the space. In Figure 14.4 and Figure 14.5, it is possible to see how regaining performance for both indexes is dependent on movement direction. Higher changes were obtained in movements performed in the ipsilateral space, where it is not possible to exploit the activation of preserved motor synergies, such as the shoulder-elbow flexion synergy: impairment in movement's recovery of the upper arm after stroke is known to be much less efficient in this part of the peri-personal workspace of patients.

EVALUATION OF EXOSKELETON-BASED TRAINING WITH A ROBOTIC-BASED ASSESSMENT

METHODS AND PARTICIPANTS

Eight patients with stroke event that occurred at least 6 months ahead (two females, six males, aged 61.33 ± 9.91) were enrolled in the treatment.

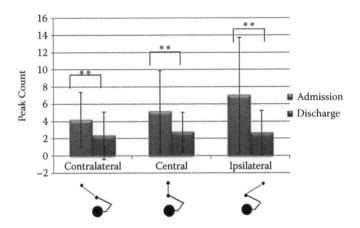

FIGURE 14.5 Smoothness index before and after training for different movement directions (**$p < 0.01$).

At the enrollment and discharge of treatment, patients underwent clinical evaluation performed with upper limb Fugl-Meyer assessment (FMA) score (66 points) and Modified Ashworth Scale (MA), on 19 muscle groups, plus a functional assessment performed with a standardized test, the BAT, consisting in evaluating in terms of time and quality of movement (estimated on a 0 to 4 numeric scale) the execution by the patient of a set of daily living tasks requiring bimanual coordination.

At the enrollment the average FMA score was 21.38 ± 11.82, and the average MA was 14 ± 9.43.

Six of the eight patients also underwent a follow-up clinical assessment 5 months after the end of treatment to evaluate the retention of regained function.

Patients enrolled in the study were administered a robotic training using the L-Exos system with training scenarios projected on a stereoscopic projection wall. The rehabilitation training consisted of two training exercises and one evaluation test, performed with the assistance of the L-Exos.

The first exercise was a reaching task performed with impedance assistance by the robot with level of assistance modulated by the robot according to the position error (impedance assistance) or with variable gain adjusted according to the force input by the patient (triggered assistance), to enhance the patient's active participation in the task.

The second training exercise consisted of a training scenario in which the patient had to compose a virtual puzzle with gravity assistance provided by the robot. The exercise was followed by an evaluation session assessing the performance in movement execution over different directions of space (Figure 14.6).

Both smoothness and time of execution of movement were measured as indexes of performance through the robot in this evaluation session. Movement time was computed as the time to move from the start position (at the center in Figure 14.6, right side) to the target at the periphery, while the SI was computed by counting the number of peaks in the velocity profile of movement.

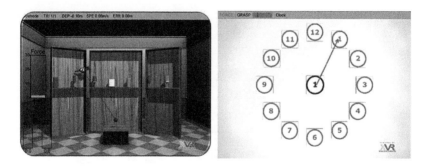

FIGURE 14.6 The rehabilitation system with the gaming scenario of reaching and the evaluation scenario.

RESULTS

We observed a significant clinical improvement in all patients. Clinical assessment at the end of treatment revealed an increase of FMA to 29.38 ± 18.42 ($p < .03$), with no significant change in spasticity (MA after treatment 10.75 ± 6.86).

On the functional level, in bimanual tasks we observed a significant reduction of time in the execution of movement from 17.06 ± 4.02 to 13.98 ± 4.48 sec ($p < .02$) with an overall increase of quality of movement from 1.99 ± 0.77 to 2.94 ± 0.95 ($p < .03$).

The assessment of the two indexes of performance performed with the robot revealed a similar trend, with a marked decrease of movement time from 2.55 ± 1.15 to 1.02 ± 0.68 sec ($p < 0.004$) and marked increase of smoothness going from 6.63 ± 0.92 to 2.75 ± 1.54 ($p < 0.002$), where decrease of the SI means increase of smoothness of movement (Figure 14.7).

We investigated the correlation of the functional indexes (bimanual activity scale) with the robot assessment scale to assess whether the improvement of performance observed with the robot was effectively transferred as ability to perform daily living activities.

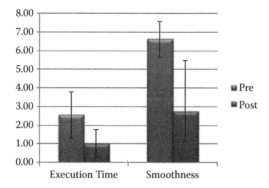

FIGURE 14.7 Assessment of movement performance with robot.

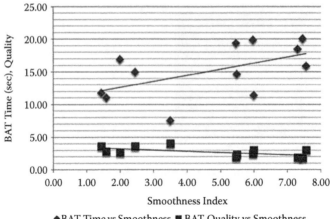

FIGURE 14.8   Correlation of bimanual activity index with smoothness of movement.

TABLE 14.1
Correlation Indexes of Bimanual Activity Scale with Robot-Based Assessment

		Execution Time	Smoothness
Bimanual time	Pearson correlation	.633(*)	.640(*)
	Sig. (two-tailed)	.027	.025
Bimanual quality	Pearson correlation	−.617(*)	−.616(*)
	Sig. (two-tailed)	.033	.033

We found a strong correlation between the SI and the time and quality of movement assessed with the bimanual activity scale, as shown in Figure 14.8 and Table 14.1. Moreover, the performance improvement in movement execution was analyzed in terms of different directions of space by means of the robot-based assessment, in terms of polar diagrams, shown in Figure 14.9. It is possible to see how a marked and significant improvement is obtained in smoothness mainly in the movements executed to reach targets in the ipsilateral space, while less significant improvement is visible in the contralateral space.

In the small subgroup of patients who underwent a follow-up evaluation at month 5 (Figure 14.10), we found a percentage improvement of FMA of 39% that was maintained also at the follow-up assessment with a percentage improvement of 27%, indicating a retention of the acquired motor skill.

DISCUSSION

Confirming our previous results obtained with kinesiologic assessment, we found that the rehabilitation training might lead to different improvements of performance in the contralateral and ipsilateral spaces.

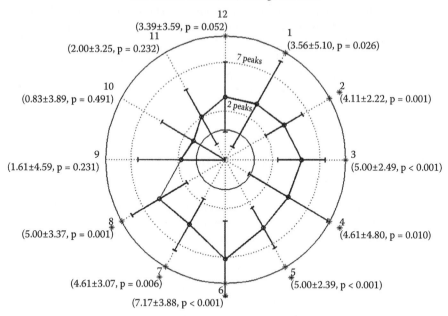

FIGURE 14.9 Percentage improvement of smoothness index along different directions of space (**$p < 0.01$; *$p < 0.05$).

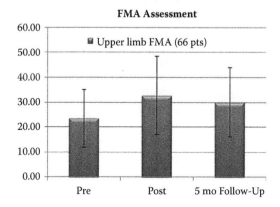

FIGURE 14.10 FMA clinical evaluation pre- and post-training and at 5 months follow-up.

Moreover, we found that the improvement in smoothness of movement represents a predictive indicator of both clinical assessment and of regained ability to perform activities of everyday life. This represents an important result, because it allows mapping directly the measured performance to improvements in terms of functional activity.

Furthermore, the first preliminary results from follow-up evaluation performed at 5 months of distance since treatment allowed us also to hypothesize that such a regaining of function is retained over time.

These data were confirmed also by analyzing through the exoskeleton, the execution time required to move virtual cubes from the central point of a circle along different radial directions, performing movements in a vertical plane. Changes in execution time and smoothness, before and after robotic training, were larger for movements performed in the ipsilateral space (Figure 14.9), which constitutes a very important goal in rehabilitation of the upper arm after stroke.

This confirms that exoskeleton systems in combination with specifically designed exercises might be used not only to enhance the process of recovery, but also to evaluate the progress of therapy.

ROBOTIC-ASSISTED BIMANUAL TRAINING AFTER STROKE WITH THE BIMANUAL HAPTIC DESKTOP

Bimanual tasks require operating the two hands together, so that they cooperate to accomplish the aimed function. In everyday life, it turns out that this cooperation is often asymmetric, with the dominant hand playing a manipulative role, when the other hand manages further stabilization (Swinnen et al., 1991). Yet, this cooperation of the two hands naturally falls into one of two types of coordination: either the simultaneous activation of homologous muscles (in-phase coordination) or alternating activation of homologous muscles (antiphase coordination). In healthy subjects, antiphase movements are less stable than in-phase movements (Kelso, 1995). The observed stability of phase or antiphase coordination seems related to the neural pathways most easily activated (Mccombe, Waller and Whitall, 2008). When acting together, a strong neural coupling exists between the two upper limbs, and this coupling differs from the neural activation in unimanual skills. At the behavioral level, this coupling is visible as a mutual influence of space-time variables of each arm in a bimanual task.

Bimanual training recently emerged in the battery of tools used in neurorehabilitation (Oujamaa et al., 2009) and now appears as one of the promising techniques in stroke rehabilitation (Cauraugh et al., 2010). Bimanual training encompasses several techniques, all of them requiring the simultaneous use of both upper limbs (Stoykov and Corcos, 2009): bilateral isokinematic training, mirror therapy using bilateral training, device-driven bilateral training, and bilateral motor priming.

Bimanual training builds on natural interlimb coordination principles to favor motor recovery (Luft et al., 2004). In a bimanual task, both upper limbs influence (and are influenced by) each other, so that it becomes possible that the nonparetic limb entrains the paretic limb and improves its output (Mccombe, Waller and Whitall, 2008). This kind of intervention can be practically implemented in a robotic device, such as the Bimanual Haptic Desktop System (BHDS) shown in Figure 14.11, where two robotic arms are used to transfer assistance from the unimpaired to the impaired limb.

The BHDS is an integrated system that merges haptic functionalities and video display terminal (VDT) systems into one. The integration of the components of BHDS has been designed to offer the best ergonomics and user comfort to the operator.

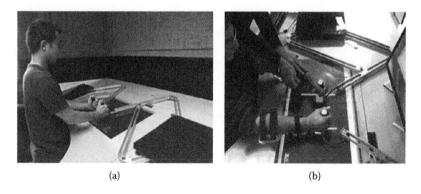

(a)   (b)

FIGURE 14.11 *(See color insert.)* Bimanual Haptic Desktop System (a) and (b) an example of one subject doing a simulated lifting task.

FIGURE 14.12 Examples of two bimanual therapy exercises with the BHDS robotic device.

Haptic functionalities are generated by two parallel planar interfaces with two DoFs each, mounted on a desk. The graphical visualization is also integrated on the desk to realize the completely coherent and colocated interaction between the force stimuli and the visual information. Two operating handles or two hand splints are mounted on the end-effecter of the device and can be used by the patients to perform a large variety of bimanual exercises, as shown in Figure 14.12. By means of the two robot arms, different control strategies can be used to transfer active force assistance on the impaired arm derived from the actin performed by the patient with the unimpaired one.

Bimanual training is also justified on neurological bases. First, the observed automatic coordination of the two arms is supported by the existence of specific neural pathways, and these pathways are activated simultaneously in both hemispheres during bimanual training. Second, brain plasticity is driven by the nature and amount of training, so that bilateral training induces reorganization in cortical networks (Luft et al., 2004).

Finally, an important functional argument favoring bimanual training is that bimanual training is closer to activities of daily living than unimanual practice: bilateral arm training seems particularly useful in the face of the high number of bimanual tasks in activities of daily living.

When used for the analysis of bimanual training in stroke patients, kinematic studies indicate that training patients with two-handed tasks improves the efficiency of grasping movements on the impaired side (Mccombe Waller et al., 2006), and that these changes are accompanied by a reorganization of brain mappings on the affected hemisphere (Luft et al., 2004).

ASSESSING MOTOR FUNCTION AFTER STROKE WITH THE BIMANUAL HAPTIC DESKTOP

Measuring recovery in stroke patients is currently done using clinical scales, such as the Fugl-Meyer score (Gladstone et al., 2002), but detailed evaluations of the quality of reaching movement with the paretic upper limb are useful tools to complement the clinical scale outcomes (Bosecker et al., 2010).

We extended this approach to bimanual manipulation, using the BHDS as a measuring tool to assess the status of patient recovery, as early as possible after admission in the rehabilitation center. In this study, we used a virtual lifting task, performed bimanually: using both hands simultaneously, the patient had to lift a box displayed in 3D on the BHDS screen, and he or she could do so by performing a synchronous outward movement of the two hands in the mid-sagittal plane. Because most early stroke patients cannot move with their impaired arm, we added a robotized help to entrain the impaired hand using the nonimpaired hand as a model. Stated differently, we used a "mirroring helping force" that acted like a spring on the impaired hand, using the nonimpaired hand as a master. This strategy allowed patients with a very limited (or no) movement capability to succeed at the task, thus making this bimanual lifting task a potentially valuable test to monitor patient recovery from the very beginning of the rehabilitation program.

We tested 11 stroke patients (9 men; mean age 59.4; 3 chronic patients; all but 1 with a severe deficit with a FMA score below 30). Inclusion criterions gathered absence of neurological or neuromuscular disorders, no orthopedic history, no pain when moving upper limb, no severe spasticity (Ashworth score ≤ 2), and no right-hemisphere syndrome (i.e., neglect).

We found a positive correlation between the movement capabilities measured with the BHDS and the FMA clinical scale. More precisely, during a lifting movement, the percentage of time spent in the upper half (T_{50}) explained 60% of the FMA score on average, with the following equation:

$$FMA = 0.15 * T_{50} + 3.22; r^2 = 0.60 ; p < 0.01$$

This result confirmed that robotized devices can be used to allow severely impaired patients to perform otherwise impossible tasks, but it especially confirmed our insight that the perfectly controlled help provided by the robot would allow us to use the results in performing the task as a measure of recovery. It is also important to point out that it was the use of a bimanual symmetric task that allowed the BHDS to provide the adequate help in real time, based on the monitoring of the movement of the nonimpaired hand that served as a master.

CONCLUSIONS

Robotic technologies offer the unique opportunity to individually tailor the physical therapy during neurorehabilitation training to the specific patient's needs, creating the conditions to enhance patient involvement in the training program, through the interplay of different factors, such as motivation, feedback, controlled assistance, and ecological presentation of exercises. Results from our studies demonstrate that robotic training can be effective both in terms of clinical efficacy and regain of motor function, assessed either by kinesiologic analysis of performance in movement execution or by functional tests in everyday life actions.

Moreover, the recording capabilities of robotic technologies make it possible to perform a continuous evaluation of motor performance through quantitative and objective indexes that correlate well with clinical scales, providing an on-line and dynamic assessment of patient's status and a record of the recovery process.

REFERENCES

Amirabdollahian, F., Loureiro, R., and Harwin, W. 2002. Minimum jerk trajectory control for rehabilitation and haptic applications. In: *IEEE International Conference on Robotics and Automation* (ICRA 2002) 11–15 May 2002, Washington DC, pp. 3380–3385.

Bosecker, C., Dipietro, L., Volpe, B., et al. 2010. Kinematic robot-based evaluation scales and clinical counterparts to measure upper limb motor performance in patients with chronic stroke. *Neurorehabilitation and Neural Repair* 24: 62.

Cameirao, M., Bermudez, I.B.S., Duarte, O.E., et al. 2009. Augmenting the rehabilitation gaming system with haptics and orthosis. Abstract: Front. Neuroeng. Conference, Annual CyberTherapy and CyberPsychology Conference 2009.

Cauraugh, J.H., Lodha, N., Naik, S.K., et al. 2010. Bilateral movement training and stroke motor recovery progress: A structured review and meta-analysis. *Human Movement Science* 29(5): 853–870.

Frisoli, A., Borelli, L., Montagner, A., et al. 2007. Arm rehabilitation with a robotic exoskeleleton in Virtual Reality. In: *IEEE ICORR, 10th International Conference on Rehabilitation Robotics*, Noordwijk, the Netherlands. IEEE, pp. 631–642.

Frisoli, A., Procopio, C., Chisari, C., et al. 2012. Positive effects of robotic exoskeleton training of upper limb reaching movements after stroke. *Journal of Neuroengineering and Rehabilitation*. In press.

Gladstone, D.J., Danells, C.J., and Black, S.E. 2002. The Fugl-Meyer assessment of motor recovery after stroke: A critical review of its measurement properties. *Neurorehabilitation and Neural Repair* 16: 232.

Hallett, M. 2001. Plasticity of the human motor cortex and recovery from stroke. *Brain Research Reviews* 36: 169–174.

Han, C.E., Arbib, M.A., and Schweighofer, N. 2008. Stroke rehabilitation reaches a threshold. *PLoS Computational Biology* 4: e1000133.

Horn, S.D., Dejong, G., Smout, R.J., et al. 2005. Stroke rehabilitation patients, practice, and outcomes: Is earlier and more aggressive therapy better? *Archives of Physical Medicine and Rehabilitation* 86: 101–114.

Kelso, J.a.S. 1995. *Dynamic Patterns. The Self-Organization of Brain and Behavior.* Cambridge, MA: MIT Press.

Kwakkel, G., Kollen, B., and Twisk, J. 2006. Impact of time on improvement of outcome after stroke. *Stroke* 37: 2348–2353.

Langhorne, P., Bernhardt, J., and Kwakkel, G. 2011. Stroke rehabilitation. *The Lancet* 377: 1693–1702.

Lo, A.C., Guarino, P.D., Richards, L.G., et al. 2010. Robot-assisted therapy for long-term upper-limb impairment after stroke. *New England Journal of Medicine* 362: 1772–1783.

Luft, A.R., Mccombe-Waller, S., Whitall, J., et al. 2004. Repetitive bilateral arm training and motor cortex activation in chronic stroke: A randomized controlled trial. *JAMA* 292: 1853–1861.

Mccombe Waller, S., Harris-Love, M., Liu, W., et al. 2006. Temporal coordination of the arms during bilateral simultaneous and sequential movements in patients with chronic hemiparesis. *Exp. Brain Res.* 168: 450–454.

Mccombe Waller, S., and Whitall, J. 2008. Bilateral arm training: Why and who benefits? *NeuroRehabilitation* 23: 29–41.

Mccrea, P.H., Eng, J.J., and Hodgson, A.J. 2002. Biomechanics of reaching: Clinical implications for individuals with acquired brain injury. *Disabil. Rehabil.* 24: 534–541.

Mehrholz, J., Platz, T., Kugler, J., et al. 2009. Electromechanical and robot-assisted arm training for improving arm function and activities of daily living after stroke. *Stroke* 40: e392–e393.

Oujamaa, L., Relave, I., Froger, J., et al. 2009. Rehabilitation of arm function after stroke. Literature review. *Ann. Phys. Rehabil. Med.* 52: 269–293.

Roby-Brami, A., Feydy, A., Combeaud, M., et al. 2003. Motor compensation and recovery for reaching in stroke patients. *Acta Neurol. Scand.* 107: 369–381.

Stoykov, M.E., and Corcos, D.M. 2009. A review of bilateral training for upper extremity hemiparesis. *Occup. Ther. Int.* 16: 190–203.

Subramanian, S.K., Yamanaka, J., Chilingaryan, G., et al. 2010. Validity of movement pattern kinematics as measures of arm motor impairment poststroke. *Stroke* 41(10): 2302–2308.

Swinnen, S.P., Young, D.E., Walter, C.B., et al. 1991. Control of asymmetrical bimanual movements. *Exp. Brain Res.* 85: 163–173.

15 Training Platforms for Industrial Maintenance and Assembly

Teresa Gutierrez, Nirit Gavish, Sabine Webel, Jorge Rodriguez, and Franco Tecchia

CONTENTS

INTRODUCTION TO INDUSTRIAL MAINTENANCE AND ASSEMBLY TASKS

Industrial maintenance and assembly (IMA) is usually a complex task that involves the knowledge of specific procedures and techniques for each machine. For example, Figure 15.1 shows several stages of the process of maintaining an electronic actuator of a motorized modulating valve by replacing worn-out parts. This process is composed of a sequence of more than 150 steps and can require over 2 hours to be completed. IMA task performance imposes high memory and knowledge demands with respect to the way the task should be performed (Neumann and Majoros, 1998), as well as competences in the movements and forces that should be applied.

FIGURE 15.1 Stages of the maintenance of an electronic actuator.

For an expert technician, the major skills involved in an IMA task are the procedural skills. Procedural skills represent the ability of the technician to obtain a good representation of how to perform each of the steps of a task as well as the correct order in which to perform them, which is captured by their hierarchic organization (Anderson, 1982; Annett, 1996; Rittle-Johnson et al., 2001). Hierarchical structures are based on both procedural memory and an adequate mental model of the machine and its components (Cañas et al., 2001; Norman, 1983, 1988). In summary, procedural skills reflect the ability of the technicians to know what actions should be performed, when to perform them, and how to perform them.

IMA tasks are common activities because industrial machinery needs periodic maintenance procedures that are usually complex and lengthy. As the variety of the produced machines is always increasing, the manufacturers have to continuously administer training to a large number of technicians to keep them updated. These training sessions have been traditionally carried out by personal class teaching, in which video documentation, two-dimensional (2D) mechanical drawings, and live demonstrations on real equipment (if available) are still the common practice. The 1-day to 1-week classes mostly involve training on assembly and disassembly procedures of company-produced machines. The result is a time-consuming and expensive training process that every industrial firm has to sustain. Therefore, improvement in the efficiency of the training process is a highly important and cost-effective challenge.

Augmented reality (AR) and virtual reality (VR) technologies may provide alternatives and improvements to traditional training methods. VR-based systems can constitute an enactive approach to skill acquisition (i.e., learning primarily by doing) through multimodal interaction with virtual scenarios, which eliminates the constraints of employing real equipment (e.g., availability, time, cost, and safety constraints), and may provide higher flexibility in task simulation. AR-based platforms can potentially improve the cognitive parts of training. AR can provide virtual instructions, information, and objects that are collocated with the real-world environment and thus both real and virtual objects can be represented in one spatially integrated scene, improving the information encoding and recall processes (e.g., Haritos and Macchiarella, 2005).

The present chapter analyzes the use of AR and VR technologies to train technicians on IMA tasks. Below, we describe the fundamental research performed in procedural skill learning, the application of the results of this research in the development of two new IMA training platforms, and the execution of a skills transfer study for the validation of both platforms.

An example of the type of IMA tasks addressed in the new training platforms is the maintenance of an electronic actuator, as shown in Figure 15.1. This maintenance task requires performing unimanual and bimanual operations, with or without using tools, on rigid and flexible components (e.g., cables). The specific procedure selected for the evaluation of the platforms was composed of four main activities: changing the level sensors of the electronic board, the belt roller, and the clamp. These activities require performing common operations in IMA tasks such as inserting/removing pieces with one or both hands (e.g., to place the electronic card in a support plate), screwing/unscrewing with a screwdriver (e.g., to fix the sensors, clamp, etc.), tightening screws with nuts and a wrench (being optional the use of other elements such as lock washers), plugging/unplugging cables, fixing cables with a cable tie, and so forth.

RESEARCH ON TRAINING APPROACHES FOR INDUSTRIAL MAINTENANCE AND ASSEMBLY (IMA) TASKS

Most researchers agree that procedural skills are developed gradually as a result of practice through repeating exposures to a certain task (e.g., Gupta and Cohen, 2002; Newell, 1991). Nevertheless, there is still a dearth of coherent recommendations for designing VR systems for procedural skills training in general, and particularly IMA activities training. In the past five years, several studies have been conducted to address four related research issues: observational learning, cognitive fidelity versus physical fidelity, guidance aids in spatial task training, and the provision of enriched information. The research described in this section provided some insights for the development of virtual platforms and training protocols (Gavish et al., 2011b).

OBSERVATIONAL LEARNING

In contemporary research, training with VR systems is frequently driven by the enactive conceptual framework that stresses the importance of direct active interaction with enriched physical environment to improve perception and enhance learning (e.g., Stoffregen et al., 2006). Nonetheless, under the theory of embodied cognition (Wilson, 2002), verbal or visual stimuli are neurologically coupled with—and can prime—related motor activities. Hence, observational learning might replace active learning during part of the training process. Two studies were conducted to address the use of observational learning in VR training systems.

The first study made use of a 75-step Lego® assembly task, where the object to be assembled was a model helicopter (Yuviler-Gavish et al., 2011a). Training was performed using a three-dimensional (3D) haptic VR system. Two conditions were compared: an active condition, in which the trainee had to both identify each target brick and position it correctly in the Lego model; and a partly observational learning condition, in which the trainee was required only to select the correct brick, which would then be automatically positioned by the system. Results showed that training time was reduced substantially with the incorporation of an observational learning phase, while performance with the real Lego test was similar for both conditions. The conclusions were that observational learning can enhance training efficiency if integrated properly within VR training.

A second study evaluated the effect of two modes of observational learning: dyad performance, in which the trainer and trainee perform the task together during the training process, and preliminary observational learning, in which the training starts with a demonstration, and the trainee performs the task only after acquiring a basic knowledge of the necessary steps. A 3D computerized puzzle task was used for this purpose (Yuviler-Gavish and Shelef, 2011). The results demonstrated that the dyad trainer-trainee performance resulted in longer training time but better performance in terms of success rates. In addition, preliminary observational learning led to shorter training time and did not influence performance. The cost-effectiveness matrix that was found in this study can assist in designing guidelines for choosing the appropriate observational learning methods for each task.

COGNITIVE FIDELITY VERSUS PHYSICAL FIDELITY

Procedural skills are acquired through repeated exposure to a certain task (Gupta and Cohen, 2002). Hence, it is important to determine whether this task should be physically or cognitively represented within the simulator. The physical fidelity approach claims that the simulator should replicate the real-world task to the greatest possible degree (Baldwin and Ford, 1988). In contrast, the cognitive fidelity approach proposes that training can be limited to focus on the cognitive activities involved in the real-world task, without needing to duplicate the physical elements of the task (Lathan et al., 2002).

In our study we compared real-world training and two alternative virtual trainers, one emphasizing the physical fidelity and the other the cognitive fidelity of the task (Hochmitz and Yuviler-Gavish, 2011). Participants were randomly assigned to one of four training groups in the 75-step Lego assembly task: Virtual-Physical Fidelity, Cognitive Fidelity, Real-World and Control. The Virtual-Physical Fidelity condition involved "building" virtually the model using a computer program. The Cognitive Fidelity training consisted of active visualization and verbalization of the different assembly stages for the same model. Participants in the Real-World training group built the model by hand. Finally, those in the Control group received no training of any kind. A post-training test was conducted to assess the development of procedural skills. Results showed that Cognitive Fidelity, but not Visual Fidelity, was inferior in terms of test time compared with the Real-World training. In contrast, Real-World and Cognitive Fidelity required less time for error correction. It seems that the visual and cognitive fidelity methods have complementary advantages, hence combining physical fidelity and cognitive training methods can enhance procedural skills acquisition.

GUIDANCE AIDS IN SPATIAL TASK TRAINING

During the execution of a task, performers may need to receive direct information about how to proceed: guidance aids. However, the addition of guidance aids may have adverse effects on performance in spatial tasks, because it encourages shallow performance strategies and may inhibit active exploration of the task. To examine this potential training trap, two experimental studies were conducted.

In the first study (Yuviler-Gavish et al., 2011b), each trainer instructed trainees on how to perform a 3D virtual puzzle under two conditions: vocal guidance, where only vocal instructions were possible; and vocal guidance with mouse pointing, where the trainer could also use a mouse to point out the target positions on the trainee's screen. The results showed that while the use of the mouse pointer reduced the mental load of the trainees during training, it also drastically lowered performance level on a nonsupervised test. These results suggest that the addition of guidance information should be considered very carefully. However, specific training protocols can eliminate the negative effect of the guidance tools. This was the research goal of the second study. This study made use of the 75-step Lego assembly task and a 3D haptic virtual reality system. Two conditions were compared: building the virtual Lego model with the help of a printed instruction book (with step-by-step diagrams of the 75 stages); and building the virtual Lego with the help of direct aids provided by the virtual platform (change the color of the target brick and then show a copy in its final position) but only on demand of the participant. Both groups showed similar performance in the real Lego test. Results showed that the use of guidance aids in a proper and controlled way, such as providing aids only on trainees' demand, does not impede task learning.

ENRICHED INFORMATION

Several past studies have demonstrated that when learners of a new procedural task are provided with enriched information about the task, in addition to the "how-to-do-it" knowledge, their performance becomes more accurate, faster, and more flexible (Kieras and Bovair, 1984; Taatgen et al., 2008). Taking into account that industrial maintenance tasks are based on complex procedures, it was hypothesized that dividing the whole task into a set of logical subtasks and providing enriched information about these subtasks could enhance the ability of the trainees to develop an appropriate and more accurate mental model of the entire task, and as a result, they could improve their performance.

A study was conducted to test this hypothesis and explore the effect of training with various levels of information. This study used a training program to teach how to assemble a 75-step Lego helicopter model. Three experimental conditions were compared: (1) step-by-step instructions and information about the entire model to be assembled, (2) step-by-step instructions and information about submodels within the entire model, and (3) only step-by-step instructions. Although the results showed similar performance for all groups, there was some evidence of benefit for training with the submodel approach (i.e., providing information at the subtask level). Therefore, the division of the entire task into a set of logical subtasks and the use of enriched information should be encouraged.

TRAINING PLATFORMS FOR INDUSTRIAL MAINTENANCE AND ASSEMBLY TASKS

Two new platforms for training maintenance tasks were developed, one based on VR technologies (IMA-VR) and the other on AR technologies (IMA-AR). These two platforms cover the different aspects and constraints of a training process:

- Using the IMA-VR system trainees learn the task within a virtual environment (VE). This platform is a suitable learning tool in situations where the real scenario is not available for training sessions, for example, when the machine is still in the design phase, or it is in use and cannot be switched off, or when errors during the training session can damage it.
- Using the IMA-AR system trainees perform assembly operations on real machines with real tools. This platform can be used as a learning tool in situations where the real system is available during training.

Both platforms and the training protocols follow the recommendations obtained from the research described above. They share the same approach for task organization (a maintenance task is defined by a sequence of steps; a set of related steps with a common goal is considered a subtask, so the whole task is divided into a set of logical subtasks). They also provide the trainees with controlled guidance aids (e.g., information about the current action) and enriched information about the target task (information at the step or at the subtask level and information about the task progress).

VIRTUAL REALITY TRAINING PLATFORM (IMA-VR)

The goal of this platform is to provide training in a controlled multimodal virtual environment for transferring motor and cognitive skills involved in assembly and maintenance tasks. It supports the enactive approach of "learning by doing" by means of an active multimodal interaction (visual, audio, and haptic) with the virtual scenario, so the trainees can interact and manipulate the components of the virtual scene, sense the collisions with the other components, and simulate different assembly and disassembly operations (see Figure 15.2). It also allows remote supervision/training through on-line interaction between trainer and trainee.

IMA-VR Setup

The IMA-VR platform consists of a screen displaying the 3D graphical scene corresponding to the maintenance task, one haptic device, and the training software to simulate and teach assembly and disassembly tasks. One of the main features of this platform is its flexibility to adapt itself to the task demands and trainees' needs/ preferences. Thus, the system can be used with different types of haptic devices (e.g.,

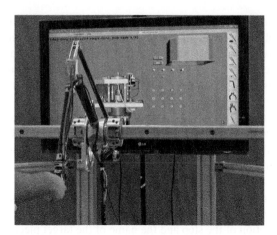

FIGURE 15.2 *(See color insert.)* IMA-VR training platform: The trainee interacts and manipulates the virtual scene combining haptic, audio, and visual feedback.

desktop or large space devices, one or two contact points, etc.). Moreover, in order to send commands to the application (e.g., to grasp/release a piece, change the view, zoom in/out, etc.), the trainees can use the keyboard or the switches of the haptic device stylus. The IMA-VR software has a modular architecture, using special APIs to increase the flexibility of the platform, and employs a multithread solution in order to provide the trainees with a consistent and proper interaction.

As shown in Figure 15.2, the 3D-graphical scene is divided into two areas. The virtual machine is rendered in the center of the scene, and the pieces that will be assembled are placed at the back wall. On the right part of the screen there is a configurable "tools menu" with the virtual tools that can be chosen to accomplish the different operations of the task. Throughout the training session, the system displays different types of information on the screen, such as information about the "task progress," technical descriptions of the components/tools, critical information about operations, and error descriptions. This critical information can also be sent through audio messages.

The platform provides different utilities to configure the training process. For example, it allows the possibility of starting the training session at any step, to select the constraint in the sequence order (the steps can follow any order or a fixed order), and to allow making steps automatically so the trainees can jump the easy steps and focus on the complex ones or even "undo steps" to repeat difficult steps. The system also automatically logs information about the task execution for further analysis of the trainee's performance.

IMA-VR Training Strategies

The IMA-VR platform provides several training strategies with different degrees of interaction between the trainees and the system. The selection of the best strategies to learn a task will depend on the profile of the trainee and the complexity of the task. A brief description of each strategy is detailed below:

- *A training strategy based on observational learning*: This is the only strategy of the platform in which the trainees do not have active interaction with the virtual scenario. The system just provides visual information about how to undertake the task in order to help trainees develop a mental model of the assembly/disassembly process. This strategy should be used at the beginning of the training process.

- *A training strategy based on "current action" aids*: This strategy provides direct information about the immediate action that trainees have to undertake, but only when they request it. The information is provided by means of (1) visual aids such as highlighting the target tool icon or changing the color of the target piece or displaying a copy of the target piece in its final position; (2) haptic aids such as exerting an attraction force toward the target piece/position; and (3) textual message aids such as displaying the name/description of the step and the name of the target piece at the bottom of the screen.

- *A training strategy based on "step-level" aids*: This strategy provides information about the current step in its totality (see Figure 15.3, on the left), on demand of the trainee. The information is provided by means of (1) visual aids such as displaying a second graphical window in which the system shows all pieces and tools involved in the current step in their final position; and (2) textual message aids such as displaying the step's name/description and a dynamic list with the names of the step's pieces/tools. If the "step-level" aid is activated and trainees still do not know how to continue with the task, they can, on demand, receive information about the immediate action up to the "current action" aid explained above.

- *A training strategy based on "subtask-level" aids*: This strategy provides information about the current subtask (see Figure 15.3, on the right) on demand of the trainee. The information is provided by means of (1) visual aids such as displaying a second graphical window in which the system shows all pieces involved in all steps of the current subtask in their

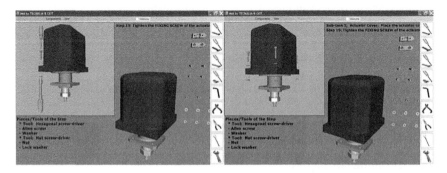

FIGURE 15.3 Left: step-level aids. A copy of all pieces/tools involved in the current step (to tighten a screw) is displayed in a second graphical window. Right: subtask-level aids. A copy of all pieces involved in the current subtask (to place and fix the actuator cover) is displayed in a second window. In both cases, a dynamic list with the names of all pieces/tools of the current step is displayed.

final position; and (2) textual message aids such as displaying the name/ description of the current subtask/step and a dynamic list with names of the pieces/tools of the current step. Similarly, as in the previous strategy, if this aid is activated but the trainees continue to require help, they can request the "current action" aid for the immediate action.

AUGMENTED REALITY TRAINING PLATFORM (IMA-AR)

The aim of the IMA-AR platform is to provide an efficient multimodal augmented reality-based training platform for improving the maintenance and assembly skills of technicians in order to accelerate their acquisition process for performing new maintenance and assembly tasks. Using AR technologies, instructions can be augmented in the trainee's field of view and directly linked to machine parts by superimposing virtual information over the live video (Webel et al., 2011).

IMA-AR Setup

Trainees interact with the platform using a tablet PC acting as an interactive see-through device. For this purpose, an additional camera is attached to the tablet PC. A movable mount positions the tablet PC while still using it as a see-through device. Thus, the trainee has both hands free for physically performing the training task. Furthermore, the trainee is equipped with a vibrotactile bracelet providing additional information through haptic feedback (see Figure 15.4, left).

The training application covers three main functionalities—the configuration of the training parameters, the execution of the training, and the generation of output— in order to analyze and assess the performance of the trainee. In the configuration menu the trainer can define individual training conditions, including enabling/ disabling of the different aids, setting the training phase, and selecting the training mode (assembly/disassembly). During training, a graphical user interface provides elements such as buttons allowing the trainee to switch steps and make requests for aid (see Figure 15.4, right). After each training cycle a log file is generated

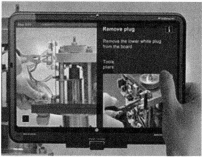

FIGURE 15.4 *(See color insert.)* IMA-AR platform. Left: The trainee interacts with the IMA-AR platform using a tablet PC that acts as a see-through device; additional haptic feedback is presented via a vibrotactile bracelet. Right: The interactive training application running on the tablet PC.

containing information about the trainee (user identification and current training phase), the application settings (configuration of aids), and the actions and performance of the trainee throughout the task execution.

IMA-AR Training Strategies

Similar to the IMA-VR platform, the AR platform pursues three training strategies, namely, the provision of *current action, step-level,* and *subtask-level* aids.

Current action aids are supportive tools that are permanently present (i.e., the trainee does not have to explicitly request them) and provide direct information about the area of interest for the next step. They are presented to the trainee in the form of 3D objects aligned with the camera image (see Figure 15.5, left).

Step-level aids pursue enhancing the information delivered by the current action aids by providing further details on demand of the trainee (i.e., detailed instructions, or abstract subliminal information that the trainee needs to interpret and transfer to the task). They provide information concerning the complete step to perform and the components involved (not only about the actions), such as hints about the required tools, screws, or washers. These aids are presented in different ways, namely, (1) additional visual objects providing details about the tools and components/pieces to be used (see Figure 15.5, left), and (2) haptic hints presenting more abstract feedback (e.g., error feedback, rotational hints indicating rotational movements, etc.).

Subtask-level aids provide information about a good mental model of the task by showing information about successive steps of the current subtask and about the condition of the device before the current step and afterwards (see Figure 15.5, right) on demand of the trainee. Because step and subtask-level aids do not strongly guide the trainee, they do not avoid the trainee's active exploration of the task, which is important for learning.

In addition to the previous aids, the IMA-AR platform presents information about the context of the step to perform (i.e., hints about the structure of the task/device and about the trainee's progress). They are presented in the form of progress bars and 3D overlays showing the structure of the device to assemble or disassemble.

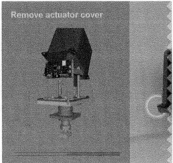

FIGURE 15.5 Left: "current action" aids (pulsing a yellow circle highlighting the area of interest for the next step) and "step-level" aids that provide detailed information about the step to perform. Right: "subtask-level" aids show information about subtasks and device conditions.

IMA SKILLS TRANSFER EVALUATION

This section presents the results of a transfer study conducted in order to evaluate the two platforms described in the previous section (Gavish et al., 2011a). The goals of this test were (1) an analysis of the efficiency of both platforms as training tools for industrial maintenance activities, and (2) a comparison between the performance obtained with these platforms and traditional training methods. The study included four experimental groups:

- Group 1—VR: participants performed the virtual task with the IMA-VR platform twice using the subtask-level-aids-based strategy
- Group 2—Control-VR: participants watched twice an instructional video demonstrating the task
- Group 3—AR: participants performed the physical task once with the IMA-AR platform
- Group 4—Control-AR: participants performed the physical task once with the help of an instructional video demonstrating the task

The selected experimental task was assembling part of an electronic actuator (presented in Figure 15.1). This task was composed of 25 steps grouped in six subtasks: place the belt roller, the two level sensors, the support plate, the electronic board, the actuator cover, and finally the clamp. The participants included 40 technicians with at least 3 years of experience in field maintenance tasks but with almost no experience in the use of VR and AR technologies. Following a between-participant design, each technician was assigned to one of the four experimental groups (10 participants in each). The demographic data of participants as well as their score in a previous capability test were similar among the groups, assuring the homogeneity of the groups.

All experimental groups followed the same protocol. In the morning, participants were trained in the task. In the afternoon, they had to perform the physical task without any help. In case they could not continue with the task, they could consult a book of pictures of the task, and this action was recorded as one "aid." At the end, the trainees filled in a questionnaire for subjective evaluation of the platforms.

Table 15.1 shows the results of the different performance measures for each group. In general, the performance of the participants in the real task was good for both training platforms, VR and AR. All technicians performed the real task without any aid, and most of them (80%) performed the task without any error; in addition, the average number of unsolved errors was low (0.3 in the AR and 0.4 in the VR platform). In the case of the AR platform, the performance time was not significantly different compared to the Control-AR group. However, the number of unsolved errors was significantly smaller ($t(18) = 2.52$, $p = 0.02$). This finding indicates that the use of the AR platform can decrease the number of unsolved errors and therefore improve technician performance. In the case of the VR platform, there were no significant differences in the final performance of VR and Control-VR groups. This was probably due to a ceiling effect with the selected experimental task after two repetitions, because the performance

TABLE 15.1
Skills Transfer Study Results (Average and Standard Deviation)

	Virtual Reality (VR)	Control-VR	Augmented Reality (AR)	Control-AR
Training Time (Minutes)	8.5 (SD = 1.3)	4.4 (SD = 0.7)	14.2 (SD = 1.9)	11.4 (SD = 2.5)
Test Time (Minutes)	8.7 (SD = 2.5)	8.7 (SD = 1.4)	8.2 (SD = 2)	8.6 (SD = 3.1)
Number of Aids	0	0	0	0
Test-Solved Errors (Number)	0.1 (SD = 0.3)	0.2 (SD = 0.6)	0.3 (SD = 0.5)	0.3 (SD = 0.5)
Test-Unsolved Errors (Number)	0.4 (SD = 1)	0.1 (SD = 0.3)	0.3 (SD = 0.7)	1.3 (SD = 1.1)

obtained by the Control-VR group was almost optimal (with a mean number of 0.1 unsolved errors).

Additionally, it is important to take into account that our volunteer technicians had almost no background in the use of VR and AR technologies. It is expected that productivity and efficiency in the use of these novel platforms will increase significantly with the familiarization of VR and AR technologies. This is particularly the case of the IMA-VR platform, which is based on a completely new interaction paradigm that requires time to use it efficiently, while interaction in the IMA-AR platform is more intuitive because it is mostly performed on real machinery.

Finally, regarding the qualitative feedback of the technicians, although the developed technologies were highly exceptional for the service technicians, most of them would recommend both platforms, VR and AR, as a training tool.

In conclusion, the skill transfer studies performed with both platforms suggest that these platforms are efficient alternatives to current training methods such as watching a video or performing the task with the support of an instruction manual for the training of technicians in IMA tasks.

REFERENCES

Anderson, J.R. 1982. Acquisition of cognitive skill. *Psychological Review* 89: 369–406.
Annett, J. 1996. On knowing how to do things: A theory of motor imagery. *Cognitive Brain Research* 3: 65–69.
Baldwin, T., and J.K. Ford. 1988. Transfer of training: A review and direction for future research. *Personnel Psychology* 41: 63–105.
Cañas, J.J., A. Antolí, and J.F. Quesada. 2001. The role of working memory on measuring mental models of physical systems. *Psicologica* 22: 25–42.
Gavish, N., T. Gutierrez, S. Webel, J. Rodríguez, M. Peveri, and U. Bockholt. 2011a. Transfer of skills evaluation for assembly and maintenance training. In B.G. Bardy, J. Lagarde, and D. Mottet. *Proceedings of the 2011 International SKILLS Conference*, 110–113. EDP Sciences.
Gavish, N., T. Gutierrez, S. Webel, J. Rodríguez, and F. Tecchia. 2011b. Design guidelines for the development of virtual reality and augmented reality training systems for maintenance and assembly tasks. In B.G. Bardy, J. Lagarde, and D. Mottet. *Proceedings of the 2011 International SKILLS Conference*, 114–117. EDP Sciences.

Gupta, P., and N.J. Cohen. 2002. Theoretical and computational analysis of skill learning, repetition priming, and procedural memory. *Psychological Review* 109: 401–448.

Haritos, T., and N.D. Macchiarella. 2005. A mobile application for augmented reality for aerospace maintenance training. *Proceedings of the 24th Digital Avionics Systems Conference, Avionics in Changing Market Place: Safe and Secure*, Washington, DC, 5.B.3-1–5.B.3-9. IEEE.

Hochmitz, I., and N. Yuviler-Gavish. 2011. Physical fidelity versus cognitive fidelity training in procedural skills acquisition. *Human Factors*. 53: 489–501.

Kieras, D.E., and S. Bovair. 1984. The role of mental model in learning to operate a device. *Cognitive Science* 8: 255–273.

Lathan, C.E., M.R. Tracey, M.M. Sebrechts, D.M. Clawson, and G.A. Higgins. 2002. Using virtual environments as training simulators: Measuring transfer. In *Handbook of Virtual Environments: Design, Implementation, and Applications*, Ed. K.M. Stanney, 403–414. Mahwah, NJ: Erlbaum.

Neumann, U., and A. Majoros. 1998. Cognitive, performance, and system issues for augmented reality applications in manufacturing and maintenance. *Proceedings of IEEE the Virtual Reality Annual International Symposium (VRAIS)*, 4–9. IEEE.

Newell, K.M. 1991. Motor skill acquisition. *Annual Review of Psychology* 42: 213–237.

Norman, D.A. 1983. Some observation on mental models. In *Mental Models*, ed. D. Gentner and A. Stevens, 7–14. Hillsdale, NJ: Erlbaum.

Norman, D.A. 1988. *The Design of Everyday Things*. New York: Doubleday/Currency.

Rittle-Johnson, B., R. Siegler, and M. Alibali. 2001. Developing conceptual understanding and procedural skill in mathematics: An iterative process. *Journal of Educational Psychology* 93: 346–362.

Stoffregen, T.A., B.G. Bardy, and B. Mantel. 2006. Affordances in the design of enactive systems. *Virtual Reality* 10: 4–10.

Taatgen, N.A., D. Huss, D. Dickison, and J.R. Anserdon. 2008. The acquisition of robust and flexible cognitive skills. *Journal of Experimental Psychology: General* 137: 548–565.

Webel, S., U. Bockholt, T. Engelke, M. Peveri, M. Olbrich, and C. Preusche. 2011. Augmented reality training for assembly and maintenance skills. In B.G. Bardy, J. Lagarde, and D. Mottet. *Proceedings of the 2011 International SKILLS Conference*, 337–380. EDP Sciences.

Wilson, M. 2002. Six views of embodied cognition. *Psychonomic Bulletin and Review* 9: 625–636.

Yuviler-Gavish, N., J. Rodríguez, T. Gutiérrez, E. Sánchez, and S. Casado. 2011a. (Revised and resubmitted). Observational learning versus enactive approach in virtual reality training. *Human Factors*.

Yuviler-Gavish, N., and M. Shelef. 2011b. (Revised and resubmitted). Evaluating two modes of observational learning in cognitive-spatial task training. *International Journal of Human-Computer Studies*.

Yuviler-Gavish, N., E. Yechiam, and A. Kallai. 2011. Learning in multimodal training: Visual guidance can be both appealing and disadvantageous in spatial tasks. *International Journal of Human-Computer Studies* 69: 113–122.

Gentle, M. and P.L. Cohen, 2002. Theoretical and computational analyses of skill training. Approaches to training-based research and management: New behavioral review, 199: 201–455.

Hanson, E. and S.D. Macchiarella, 2005. A holistic approach for the augmented reality for aerospace maintenance training. Proceedings of the 24th Digital Avionics Systems. Conference in Chicago, Morgan Kaufmann. See end-access, Washington, 1-4, pp: A.B.2-1-A.B.2-1-10.

Pischin-Ruben, I. and K. Tucker, Gerett, 2011. Physical reality versus cognitive reality review in procedural skills acquisition, Graphic research 2.2a, 39: 305–320.

Maran, J.P. and S. Gorman, 1999. The role of mental model in learning to operate a device. International Science 9, 13: 35–72.

Langdon, A., M.H. Peters, M.M. Schooling, D.W. Carnemeret, C.A. Hingam, 2011. Using virtual environments in memory therapies: Using data-centric 11-12 interfaces, A project on Neuropsychology and Rehabilitation, conference proceeding of C.I.E.M.I. Manchester, pp: 112-116, ICM, and Detroit.

Merciament, Loon, A., Morer, et al., 1997. A primer of appraisal and assessment for presentation and simulation in health settings in American Journal of Medicine, 103: 1-25-116.

Schuck, P. and O. Lewis, 2006. Maintenance based mental models: In the age of technology. IT learning, 42(6): 33-b, 1-45. Springer.

16 Training Approaches for Improving Robot Programming-by-Demonstration Skills

Volker Schmirgel, Uwe Zimmermann,
Eldad Yechiam, Ariel Telpaz, Thomas Hulin,
and Carsten Preusche

CONTENTS

INTRODUCTION

With the KUKA-DLR light-weight robot (LWR) a new generation of robot technology became available to the market (Hirzinger et al., 2002). This robot has several new innovative features that differ from traditional industrial robotics. It can lift up to 7 kg with full dynamics, even though the robot only weighs 15 kg. With its seven axes it is a kinematically redundant robot that provides high flexibility similar to the human arm. But most important, due to its internal sensors, the robot can react to external forces all along its structure. On the one hand this allows for sophisticated impedance control that enables the user to parameterize the compliance of the robot system and how it reacts to external forces. Furthermore it is possible to apply specific forces and torques to the environment with impedance control. On the other hand it allows users to manually guide the robot in an easy and intuitive way and enables *programming by demonstration* (PbD) (Billard et al., 2008).

PbD is a programming paradigm (Koeppe, 2001) with the aim to avoid the cumbersome writing of a robot program as it is done in traditional programming (Biggs and MacDonald, 2003). The worker who knows how to solve a task but who is not an expert in robot programming should be enabled to just "demonstrate" the task to the robot by taking the robot by the hand. PbD should not be confused with programming by imitation (Maeda et al., 2002; Pardowitz et al., 2007), where the user performs a task without a robot and is observed in such a way as to automatically generate a robot program that executes the same task.

In PbD the human is in direct physical interaction with the robot system. This also means that the user needs to know some robot-specific knowledge. He does not necessarily need to know how to program a robot but he has to understand the behavior and reaction of the robot system. Examples for this kind of knowledge include

- Kinematic specific issues (e.g., workspace, redundancy, singularities, joint limits, etc.)
- Compliant behavior
- Dynamic restrictions (e.g., maximal velocities, forces, etc.)
- Coordinate systems

By introducing PbD and the new LWR features, the focus of skills to be learned has shifted (Schmirgel et al., 2008). With traditional programming users had to learn typical programming issues (e.g., control-structures like "if...then," variables, and commands). This kind of knowledge could be learned by textbooks, lectures, and exercises. With PbD it is more important to understand the system behavior. Similar to learning how to ride a bicycle, PbD is mainly learned by experience when working and physically interacting with the system. In fact, it is almost impossible to learn to ride a bicycle by reading a book about the physics behind it. The main goal of the current chapter is to describe the development and evaluation of two training systems for enhancing skill acquisition in PbD. Therefore, a typical robot application in the context of PbD was analyzed and two tasks and the related skills were identified.

The first one is a "pick and place" task, with the focus on robot motion from one position and orientation in Cartesian space to another. The specific skill the user needs is to avoid robot singularities. A robot singularity describes a situation in which a robot arm cannot be moved in certain directions anymore. For PbD singularities are problematic not only due to those restricted movements, but also to the fact that close to singularities a robot can behave unexpectedly with quick joint movements.

The second one is a "peg in hole" task. Here the focus is on setting the compliance parameters of the robot in such a way that they will be optimal to the specific task at hand. Similar to a metal spring, setting robot compliance comprises two parameters, the compliance (inverse of stiffness) and the equilibrium point. For those two training areas we have developed two training platforms, each with a different training approach and technical focus (Hulin et al., 2010), but both having the common ground of adding additional sensory information. It has very often been claimed that a combination of different sensory inputs enhances performance, particularly if they provide information that is not redundant (i.e., not identical) (e.g., van Merrienboer and Sweller, 2005).

The first training platform used virtual reality (VR) technology for training how to avoid robot singularities in a "pick and place" task (see Figure 16.1, left). It was

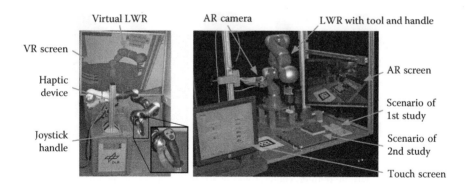

FIGURE 16.1 *(See color insert.)* The two training setups. Singularity training setup (left) and compliance training setup (right).

composed of a LWR for haptic interaction, and a three-dimensional (3D) back-projection system, showing a virtual LWR. Thus, participants perceived both stereo visualization and haptic feedback. The haptic LWR had a joystick handle as a tool, which was equipped with a three-state switch and a dead-man button. The dead-man button is a fail-safe device that stops the robot in case the button is released. By pressing that button, the user can activate the haptic robot and with it move the virtual LWR. Due to the force feedback of the haptic robot, the user can feel the effect of singularities of the virtual robot. In this system we examined the benefit due to additional visual information concerning the correct paths of avoiding singularities, in addition to the force feedback.

The second training platform was for compliance training. The hardware layout for this system comprised a LWR setup in a working cell similar to a real robot cell, with task-related equipment (e.g., gripper, handle, defined workspace, springs, or Lego® structures) (see Figure 16.1, right). Additional information, interaction facility, and task simplification was provided by a touch screen and an AR screen. The training task involved the deployment of a real robot for performing realistic robot tasks. With this setup virtual (visual and haptic) information could be augmented to real-life scenarios. In this system we examined the influence of additional haptic information beyond the numeric (visually presented) compliance information.

STUDIES ON SINGULARITIES

Our first study investigated the problem of how training for avoiding robot singularities can be optimized. For a complex robotic system with redundant kinematics (e.g., robots that have an elbow, such as the LWR), several singularities can exist with coupled conditions for joint positions, which are difficult to learn for robot users. The interesting question investigated in the present section is: How can we provide an effective training for avoiding robot singularities?

To this end, the present section introduces two kinds of visual training accelerators for the problem of robot singularities and presents an evaluation study on this accelerator. Both accelerators are visual pointers. The first kind of visual pointer (see

FIGURE 16.2 The light-weight robot (LWR) in a singular situation, and the different kinds of visual pointers; left: no pointer; center: rotating arrows; right: transparent robot.

Figure 16.2, center) is a rotating arrow that is indicating for each joint how close it is to the next singularity, and how to move out of that singularity. The second kind (see Figure 16.2, right) provides the same information but in a different manner (i.e., as an animated transparent robot, which is showing how to move the robot elbow out of singularities). This second visual pointer is only applicable for robots with redundant kinematics, whereas the rotating arrow can be applied for any robot with revolute joints.

The evaluation study on these two training accelerators was performed on the virtual reality training system, which was introduced in the previous section. In the evaluation study a virtual scenario was used, which main component is a virtual robot arm, whose singularities should be avoided. The used virtual robot was a LWR, which means that the same robot was used as a haptic device and in the virtual world. The task in the evaluation study was a pick-and-place task, which is typical for PbD. This task was embedded in a virtual Lego game, whose goal was to build a predefined Lego structure composed of five Lego bricks. The picking and placing positions were given in such a way, that the direct path would drive the virtual robot through a singularity. Therefore, to avoid singularities a longer path needs to be chosen.

In the training procedure two different Lego games had to be played, one in a training phase, and a second as a transfer task. Training started with introductory explanations and watching a video of the Lego game. Also each participant was allowed to play around with the haptic system and the virtual robot in order to get used to it. Subsequently, two Lego games were played. For each Lego game we measured the total duration until the five Lego bricks were placed correctly, and the performance with respect to avoiding singularities. This performance value is a measure for the distance to the singularities.

The participants were randomly split into four groups with at least 10 participants in each group. In the training phase, each group had to play a Lego game with different kinds of visual aids and interaction possibilities (see Figure 16.3). In contrast to the other groups, group D was allowed to move the elbow of the virtual robot by using the three-state switch of the joystick handle.

Group D was expected to perform best during training and during transfer, because it had the possibility of moving the virtual elbow.

The results of the evaluation study showed that visual feedback during training enhanced performance on the transfer. Also, as seen in Figure 16.4, group D

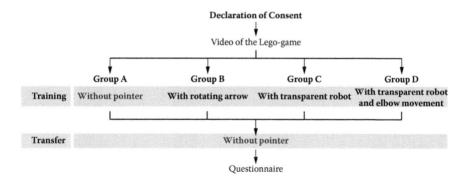

FIGURE 16.3 The training procedure with four different groups of participants.

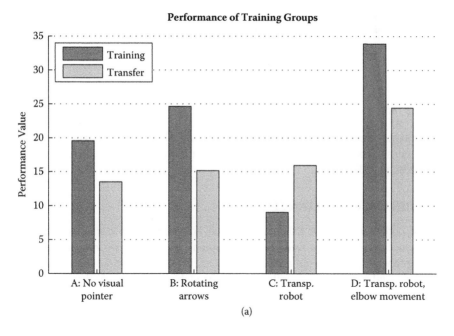

(a)

FIGURE 16.4 The evaluation results during training and transfer for the four different groups.

performed better on the transfer task than the other groups, although it took them more time to complete the training task.

We can conclude from this pattern of results that group D, which, in addition to visual feedback during training, had the possibility to manipulate the virtual robot's elbow, learned best how to avoid singularities. On the one hand, they invested more time in training, and on the other hand training became more efficient in avoiding singularities.

As for the effect of visual feedback on performance, the mere presence of visual guidance during training was by itself sufficient to facilitate performance (see Figure 16.4). Our explanation for this finding is that visual feedback on user's

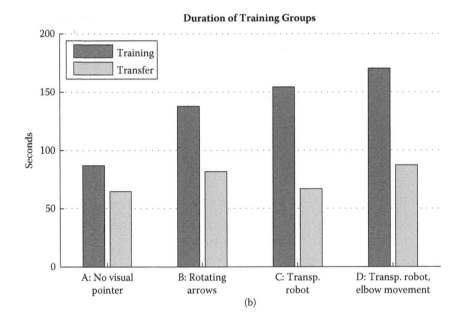

FIGURE 16.4   (continued).

performance can benefit training, rather than visual feedback on action-specific instructions, which has been proven to impair training (see Chapter 15, or Gavish and Yechiam, 2010).

STUDIES ON COMPLIANCE

Our second main field of study investigated how training to parameterize robot compliances can be improved. Robot compliances are for two reasons difficult to comprehend: First, compliance is not an attribute prevalent when dealing with machines, and second it is visually not apprehendable. Usually only the result of the compliance due to an action can be observed. For this reason using a different and adequate feedback channel promises to result in better training results.

The studies on compliance were carried out focusing on a haptic accelerator. In the current system, the user is provided with direct feedback of the parameterized levels of compliance, in the way that he or she may touch the robot's tool and directly feel and observe the compliance of the robot due to the forces intentionally applied by the user's hand. This multidimensional haptic aid is expected to support the acquisition of the cognitive model of compliance through direct haptic feedback in the course of the robot movement.

The evaluation study on the haptic training accelerator was performed on the augmented reality training system, which was introduced at the beginning of this chapter (see Figure 16.1, right). For compliant robot motion in Cartesian space, there are six compliance parameters that define the translational and rotational compliance of a given point of the robot. This is usually the tool tip. Together with a spatial difference between actual position and targeted position (which can also be described by six parameters), the compliance creates a force. In order to observe the influence of additional sensory information only, studies with reduced complexity (reduced number of parameters) were carried out.

In a research study, the task was simplified to a one-dimensional problem in Cartesian space: The robot had to press down on a spring and compress the spring by a specific distance. To reach this goal an appropriate parameter value had to be found. It was either the compliance value for a fixed but unknown target position or the target position in one direction for a fixed but unknown compliance. These two tasks were repeated with three different springs resulting in six training tasks in total, which took about 30 minutes.

The parameterization started at a default value for each task which was the same for each participant. The participants were visually observing the robot movement and the outcome of the trial, usually the robot pressed the spring too much or not enough. Then the user had to change the parameterization value with +/– buttons on the touch screen (see Figure 16.5). Step ranges were 25 N/m for compliance parameterization tasks and 5 mm for target position tasks. After having finished the parameterization, each trial was completed by starting the robot motion again and watching the result of the new parameterization.

The training tasks were performed by two groups of subjects under different conditions (see Figure 16.6). Group A was allowed to use additional haptic feedback, and Group B as the control group had to rely on visual feedback only. The participants were not allowed to touch the springs. Subjects of group A were allowed to experience the result of their parameterization by grasping the handle and exerting forces on the parameterized compliant robot. They could subsequently alter the parameters and feel the result again. With this feedback the users started robot motion when they thought that the task could be accomplished with the compliant behavior they have felt with their own hands.

For the transfer task the difficulty level was increased by mounting the springs to a fragile structure. Again the springs had to be pressed to a certain extent. The main difficulty for the subjects was that they had to set both parameters for compliance and target position (which affect each other) at the same time now. This way the subjects could show how well they understood what they had done before and how well they could use this understanding for combining their experience. For the transfer tasks all subjects were only allowed to get visual feedback of their actions.

FIGURE 16.5 Main task steps: Setting parameters on touch screen (left), haptic feedback by touch (center), and executing (i.e., compressing spring through parameterized compliance or target point) (right).

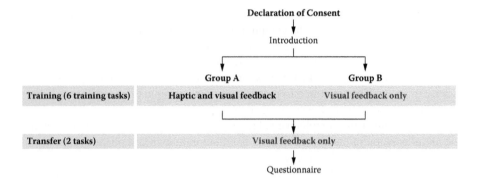

FIGURE 16.6 Training procedure for evaluating haptic feedback for parameterizing compliance.

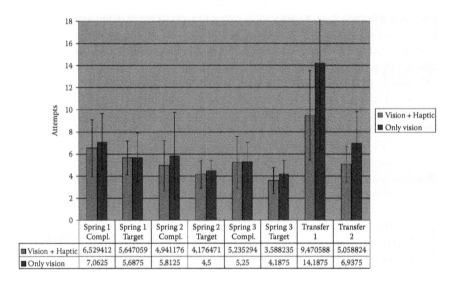

	Spring 1 Compl.	Spring 1 Target	Spring 2 Compl.	Spring 2 Target	Spring 3 Compl.	Spring 3 Target	Transfer 1	Transfer 2
■ Vision + Haptic	6,529412	5,647059	4,941176	4,176471	5,235294	3,588235	9,470588	5,058824
■ Only vision	7,0625	5,6875	5,8125	4,5	5,25	4,1875	14,1875	6,9375

FIGURE 16.7 The performance comparison for the two subject groups. Training tasks (left): setting compliance parameters (Compl.) and setting target position (Target) at three different springs. Transfer tasks (right): setting both parameters at two different springs.

By comparing the number of attempts each group needed for each task (see Figure 16.7), it becomes obvious that both groups perform similarly during training. The numbers of attempts are close together and the learning gradient for both tasks did not differ between the two groups.

It is also observable from the transfer task data that the learning process was not complete at that time. When comparing the average number of attempts in the transfer tasks, it shows that the haptic accelerator in the training phase significantly improved the transfer task performance, in which it was not used anymore. One explanation is that the haptic feedback was stimulating the awareness of what was done in the training phase. Variance in only the vision group shows that some subjects benefited to a similar extent even from the inferior feedback.

We can conclude from this study that haptic feedback is an essential information channel when training to parameterize force-based machine features like compliance control in robotics.

IMPLICATIONS TO PROGRAMMING BY DEMONSTRATION (PbD) AND ROBOTICS

By introducing new robot concepts like PbD to the broad market, new training methods are also needed for customers using such robot systems. The main things a trainee has to learn within robotics, especially when interacting physically with the robot, are mathematical and physical concepts and models, like singularities and compliance. These mental models are not easily acquired by traditional learning (e.g., textbooks and lectures). Instead enactive learning seems a far better method. The results from the technical developments and the experiments carried out have shown that by using accelerators the enactive learning can be further optimized.

This will change the way robot training will be performed in future. On the one hand there will be a shift from traditional learning to enactive learning by introducing VR training platforms at the robot manufacturer's training facilities (training center setup). On the other hand the compliance training setup has shown that the robot could be used as an interactive training device that could be directly used for training at the customer's site (on-site training).

Furthermore, smaller experiments not described in this chapter have shown that the proposed accelerators, especially the visual clues, could be used to train other robotic features, like joint limits, speed limits, redundancy, and workspace (Evreinov et al., 2011).

The new robotic features and the PbD paradigm have a positive impact on the quality of robot programs, the time needed to program a robot, and the use of robots in small and medium enterprises where no robot expert is available. But these new benefits will only be successful when users will be able to learn the needed skills in a fast and easy way. The presented work shows a possible way how to achieve these challenging requirements.

GENERAL CONCLUSIONS TO SKILLS LEARNING

The results of the studies on compliance and singularities avoidance have two main implications for multimodal skill learning and acquisition. The first implication concerns the effect of using visual aids during training. There are inconsistent findings regarding the utility of various visual aids in the course of training (e.g., Cherubini and Dillenbourg, 2007; Fussell et al., 2000, 2004; Gavish and Yechiam, 2010; Ranjan et al., 2006; Yuviler-Gavish et al., 2011). Some authors have reported positive effects of visual aids on online performance but did not examine nonsupervised training (e.g., Fussell et al., 2000, 2004; Ranjan et al., 2006). Yet others have shown that there was a negative effect of visual aids on nonsupervised performance (Cherubini and Dillenbourg, 2007; Yuviler-Gavish et al., 2011). Yuviler-Gavish et al. (2011) suggested that the negative effect is due to overguidance of the trainees afforded by visual aids such as demonstration of movement paths. Yuviler-Gavish et al. (2011) further showed that visual aids were attractive to trainees despite their negative effect on performance, a pattern they refer to as a "training trap."

The current findings show that visual guidance in the context of direction to avoid singularities was successful and helpful even when the test involved nonsupervised performance. We believe that the reason for this difference is that our design of visual guidance was deliberately constructed so as to avoid "mini-directions" of specific motor gestures and trajectories. Rather, the visual aid informed participants of an overall alternative strategy when their chosen strategy for performing the task proved to be detrimental.

The second implication for multimodal training comes from the studies on compliance. We have shown that augmenting training with haptic feedback was helpful compared to a visual-only condition. We believe that the success of haptic feedback in the current setting could also be in some sense associated with its effect on the reward structure of the task. Whereas visual instructions draw participants to focus on the specific subcommands (e.g., focusing on visual targets), haptic feedback, particularly in the context of a novel robotic setting, encourages participants to explore. Of course, there are alternative theories suggesting that mere multimodal interfaces improve performance (e.g., van Merrienboer and Sweller, 2005), but these cannot account for the interacting effect of multimodal feedback, such as that found by Yuviler-Gavish et al. (2011) and for the effect of implementation technique on the efficiency of visual guidance as evidenced by the distinct effects of performance of the different versions of the visual guidance that we implemented.

REFERENCES

Biggs, G., and MacDonald, B. 2003. Survey of robot programming systems. *Proceedings of the Australian Conference on Robotics and Automation*, CSIRO, Brisbane, Australia.

Billard, A., Calinon, S., Dillmann, R., and Schaal, S. 2008. Robot programming by demonstration. In *Springer Handbook on Robotics,* ed. B. Siciliano and O. Khatib, 1371–1394. New York: Springer.

Cherubini, M., and Dillenbourg, P. 2007. The effects of explicit referencing in distance problem solving over shared maps. *Proceedings of the 2007 International ACM Conference on Supporting Group Work*. Florida: ACM, 331–340, ACM, NY.

Evreinov, G., Raisamo, R., Hulin, T., and Zimmerman, U. 2011. PbD: Visual guidance for robot-arm manipulation, *BIO Web of Conferences*, Vol. 1, B.G. Bardy, J. Lagarde, and D. Moffat (Eds.), Montpellier, France, December 15–16.

Fussell, S.R., Kraut, R.E., and Siegel, J. 2000. Coordination of communication: Effects of shared visual context on collaborative work. *Proceedings of the CSCW 2000 Conference on Computer Supported Cooperative Work* (21–30). New York: ACM, L. Erlbaum.

Fussell, S.R., Setlock, L.D., Yang, J., Ou, J., Mauer, E., and Kramer, D.I. 2004. Gestures over video streams to support remote collaboration on physical tasks. *Human-Computer Interaction,* 19, 273–309.

Gavish, N., and Yechiam, E. 2010. The disadvantageous but appealing use of visual guidance in procedural skills training. In D. Kaber and G. Boy (Eds.), *Advances in Cognitive Ergonomics* (764–773). Boca Raton, FL: CRC Press.

Hirzinger, G., Sporer, N., Albu-Schäffer, A. et al. 2002. DLR's torque-controlled light weight robot III—Are we reaching the technological limits now? *IEEE International Conference on Robotics and Automation,* (1710–1716). Washington, DC.

Hulin, T., Schmirgel, V., Yechiam, E., Zimmermann, U.E., Preusche, C., and Pöhler, G. 2010. Evaluating exemplary training accelerators for programming-by-demonstration. *Proceedings of ROMAN IEEE 2010* (467–472), Viareggio, Italy.

Koeppe, R. 2001. *Robot compliant motion based on human skill.* Ph.D. dissertation: ETH Zürich.

Maeda, Y., Ishido, N., Kikuchi, H., and Arai, T. 2002. Teaching of grasp/graspless manipulation for industrial robots by human demonstration. *Proceedings of the IEEE/RSJ International Conference on Intelligent Robots and Systems, IROS 2002* (1523–1528), Lausanne, Switzerland.

Pardowitz, M., Glaser, B., and Dillmann, R. 2007. Learning repetitive robot programs from demonstrations using version space algebra. In *Proceedings of the 13th IASTED International Conference on Robotics and Applications*, RA 2007, Wurzburg, Germany. Schilling, K. (Ed.), 394–399, ACTA Press, Anaheim, CA.

Ranjan, A., Birnholtz, J.P., and Balakrishnan, R. 2006. An exploratory analysis of partner action and camera control in a video-mediated collaborative task. *Proceedings of the CSCW 2006 Conference on Computer Supported Cooperative Work* (403–412). Banff, Alberta, Canada: ACM.

Schmirgel, V., Zimmermann, U.E., Hulin, T., and Preusche, C. 2008. Position Paper: Human Skills for Programming-by-Demonstration of Robots. *In Beyond Movement*, 144–167. Alinea editrice s.r.l.

van Merrienboer, J.J.G., and Sweller, J. 2005. Cognitive load theory and complex learning: Recent development and future directions. *Educational Psychology Review,* 17, 147–177.

Yuviler-Gavish, N., Yechiam, E., and Kallai, A. 2011. Visual guidance can be both appealing and disadvantageous in spatial tasks. *International Journal of Human Computer Studies*, 3, 113–122.

Section V

Prospects of Multimodal Virtual Stimulators in the Training and Transfer of Skills

17 Skill Training in Virtual Reality Environments

Its Requirements and Challenges

Daniel Gopher

CONTENTS

INTRODUCTION

Chapters 2 and 3 introduced a conceptual framework and building blocks for the development of multimodal virtual reality (VR) training environments. To recapitulate, a clear distinction was made between the existing calls and research efforts in development of virtual worlds to those associated with VR training environments. The success and validity of contemporary efforts to develop virtual worlds are most commonly expressed and examined by measures of the experience of *presence* and performer's *immersion* in the virtual world (Sanchez-Vives and Slater, 2005). In contrast, the guiding constructs for the development of skill training platforms are *relevance, facilitation,* and *transferability. Relevance* refers to the ability of the VR platform to provide relevant experience for the development of competency and skill for the performance of a targeted task. *Facilitation* is manifested in the introduction of facilitation and guidance to assist and accelerate the acquisition of the designated skill. *Transferability* emphasizes the transfer of knowledge, competencies, and skill levels from VR training to the performance of the actual task in the real world.

The difference between the leading concepts in the construction of virtual worlds to those of developing VR training platforms represents their differential emphasis and originating goals. The prime goal in the effort to develop a virtual world is the creation of a virtual analogue to real-world experience, such that performers will feel and behave as close as possible to their behavior in the real world. Accordingly,

Presence is considered as the propensity of people to respond to virtually generated sensory data as if they were real (Sanchez-Vives & Slater, 2005).... This encompasses their ability to act within the environment created by virtually generated sense data in a manner commensurate with how they would be able to behave if the sensory data were real. (Slater et al., 2009)

When skill training is the prime goal, the focus is on the development of an alternative environment that will enable performers to practice and prepare for their role in the real world. These distinct and different objectives are mapped into corresponding scientific questions and engineering challenges (Gopher et al., 2010). The first 10 chapters of the book have examined in detail the different aspects of developing multimodal VR training platforms for perceptual motor skills. Chapters 11 through 16 describe six case studies in which the approach was applied to the development and testing of training platforms for tasks, in the three domains of sport and entertainment, medicine and rehabilitation, and industrial work. This chapter presents an overview of these efforts and considers possible generalizations from six case studies, in light of the general conceptual framework and leading constructs that guided their development and testing. Such a discussion brings forward major distinguishing factors and underlines the challenges that signify the development of multimodal VR platforms for skill training. This discussion examines the format and mode in which the constructs of *relevance, facilitation,* and *transfer* were defined, engineered, and experimented across the six tasks that were studied. The three constructs can be briefly described as follows:

1. *Relevance*: The ability to create part task representations that are relevant to task performance and skill training, rather than full and complete capturing and representation of the represented real-world task. Also required is a decision about the targeted population and training entry point.
2. *Facilitation*: An effort to facilitate learning through specially designed feedback and knowledge of results, as well as the selection and introduction of training scenarios in the construction of virtual worlds, which may even differ or violate normal physics.
3. *Transfer of training*: Two notions of transfer: (1) the viability of VR training as a replacement and valid alternative to real-world task training; and (2) additional benefits and value of VR training over and beyond real-world training.

RELEVANCE

When developing a training platform for a targeted task, what are the skill and task components that would be captured and represented in the virtual training environment? What situations and training scenarios should be created? Who are the targeted trainees and at what stage of their regular skill acquisition process will the virtual training platform be introduced? These are three key questions to be considered when deciding upon the format, content, and context of a developed training system. Together they represent the different aspects of the relevance of the experience acquired in the training system for the performance of the actual tasks in their operational environment. When addressing the six tasks for which training platforms were developed in the SKILLS project (surgery, competitive rowing, entertainment juggling, upper-limb rehabilitation, programming by demonstration, industrial maintenance application), it is clear that each of them is a complex and multidimensional task that cannot be completely captured and fully represented in the virtual training environment. It is evident from the review of Chapters 11 through 16 that all tasks are rich, multidimensional, complex, and dynamic such that the creation of a complete and full task VR representation is impossible and not practical. Hence, a key requirement is to determine which of the task and skill components to focus upon. Embedded in this decision are also a selection of the intended population of trainees and a consideration of the point at which virtual training should be introduced. These are key decisions because on the one hand different levels of trainees may vary at the entry point in their prior knowledge and competencies, hence differ in their preparedness and needs for training (think about rowers, surgeons, technicians, or robot operators). On the other hand, skill acquisition and training on the VR platform should feed forward to improve the ability of these trainees in their next phase of actual task performance. In this context it is also important to reiterate the special focus of the SKILLS project on the development of perceptual motor skills as well as on the value of multimodal training with an emphasis on haptics. These are also the aspects of relevance that have led engineering and software development efforts associated with each of the six task platforms. An important reference and anchor point in this context has been the unified framework of sensory-motor and cognitive subskills which has been presented in Chapters 2 and 3. The following is a brief summary of the selected task components and skills focus in each of the six training platforms:

Rowing: The training platform for rowing has three focal points: Development of basic rowing skills; control of effort and energy management; improving coordination in team rowing. Training of basic rowing skills focused on technique optimization directed to provide intermediate rowers with a correct representation of the rowing cycle to properly start and maintain rowing. Effort control and energy management targeted the best exploitation of resources during a race. Coordination and synchronization with team members is a key competency for successful competitive rowing. Note that each of these composites of rowing expertise combines motor, perceptual, cognitive monitoring, and executive control components. The selected trainee population has been intermediate-level rowers. This is a level at which rowers already have some experience in rowing and basic competencies to build upon in training but have not yet reached a high level, deep-rooted competencies and performance routines, which are difficult to reshape and change.

Juggling: The focus for the Light Weight Juggler (LWJ) platform has been attention management, spatial temporal relations (dwell ratio), and rhythmic bimanual coordination. Performance of a juggling trick requires performers to execute and monitor the concurrent tossing and catching of multiple objects (e.g., balls, cones) in consecutive cycles that follow a predetermined pattern. Juggling skill can be conceptualized as being made up of three primary components: pattern memory—the ability to identify, represent, and memorize the pattern and sequence of objects in specific tricks; acquisition and stable representation of the spatial-temporal relationship of catching and throwing objects to allow cyclic rhythmic performance; and motor competency of tossing and catching juggled objects. Motor competencies of the type included in real balls juggling are not represented in the investigated LWJ platform, in which only virtual balls are juggled by virtual hand palms, and a VR haptic interface is not included.

In the VR juggling trainer, the skill composites of the attention management of concurrent catching and throwing of multiple balls, maintaining the proper dwell ratio between the time balls are in hand and in the air, as well as the development of bimanual coordination of ball movements can be properly presented and trained. Similar to all training platforms, but most clearly assessed and evaluated in the LWJ platform, is the training value of part task format of the VR representation when an important skill dimension is missing (motor competency and gravitational vectors). The accompanying immediate questions are the ability to properly isolate and train part task skills, as well as the value and benefit of such training to the complete task performance in transfer to real life. The juggling platform is also the only one in the SKILLS project which was developed for naive trainees. Nonetheless, it should be recognized that the term *naive* assumes the existence in the novice's repertoire of basic motor control, bimanual hand coordination abilities, and perceptual and attention capabilities, that will allow trainees to efficiently respond and practice in the LWJ.

Maxillofacial surgery: The training platform developed for maxillofacial jaw surgery is an interesting example of skill *relevance* instantiated in a multimodal VR trainer. The platform focuses on the Epker osteotomy surgery in which basic haptic skills are of primary importance in the actual conduct of a surgery. This type of surgery requires highly controlled drillings by the surgeon with a major reliance on haptic information and limited visual information. Within this context, a training protocol has been developed with an emphasis on haptic-dependent competencies in jaw drilling procedures. The emphasis has been on the acquisition of effective and accurate competencies in the operation and use of surgery drillers and accompanying tools to perform tasks that are similar in nature and context to real surgery. This type of relevant and highly important proficiencies is most commonly not included in regular medical training programs. The skills are normally developed in the course of assisting and conducting supervised surgeries on real patients.

Training in which priority is given to the way surgery tools are operated and sensory-motor features of sound, vibration, stiffness, and vision can be provided and systematically manipulated, brings forward a broader notion of relevance of VR training platforms over and beyond those of traditional medical training, as the focus is on aspects of skills that are important but not normally included in preparatory training. The targeted population for this platform is surgeons preparing for Epker osteotomy and similar surgeries.

Industrial maintenance and assembly: This platform has been developed to enable training of technicians in multistep assembly and maintenance tasks. The trainees are technicians who already have technical knowledge and some experience with the systems on which they are called upon. Unlike the former three platforms where the focus of training was on the acquisition of perceptual motor competencies, in the present platform these skills subserve the ability to acquire and execute long, multistep procedures to be executed completely from start to end, in the right order, accurately and efficiently. The challenge has been to construct platforms and develop training programs that facilitate development of the proper encoding retention and execution of procedures, while still maintaining in the VR and augmented reality (AR) systems the key features necessary for mapping and linking task performance in the virtual to the real work environments.

Programming by demonstration: The aim of the two training platforms for the programming by demonstration task has been to improve the communication and dialogue between the human operator and the robot during the task in which the human demonstrates to the robot and takes it through the specific job that it is required to perform. Two aspects were selected for training: (1) dynamic identification of robot motion constraints to avoid singularities in which robot motion is blocked; and (2) the ability to communicate efficiently to the robot compliance levels of performing task segments and their adjustment when moving through segments. Singularity and compliance call for improved anticipation of dangers and increased sensitivity to changes and adjustments. This is a special case in which robot operators are trained to recognize and adjust their motor interaction with the robot to its limitations and information needs.

Upper-limb rehabilitation: A unique perspective on the relevance construct is presented by the application of robotic and VR technologies to expend therapeutic options for upper-limb rehabilitation. These platforms have been developed to provide tools and opportunities to assist a therapist and increase his or her power and degrees of freedom in the development of rehabilitation protocols as well as in assessing functionalities and progress of patients. Because of the considerable variability among patients in the exact nature and level of their impairment, the robotic systems and the VR task environments should provide focused modes of interaction that can be tailored and customized by the therapist for the needs of a treated patient. Accordingly, the exoskeleton system (EXO) enables the execution of single-hand 3D movements that can be tracked and if desired constrained by joints. Movements can be performed by the patient or the robot alone or may be combined and share responsibility in any proportional ratio. The bimanual coordination system (BMS) is designed for tasks that require the joint, coordinated use of the two hands. Accordingly, tasks for the exoskeleton require pointing, aiming, reaching, performing circular motions, rearranging, and building blocks. For the bimanual system they focus on lifting, relocating, and towering of objects with both hands. The exact way in which tasks are selected, ordered, and scheduled to form a rehabilitation protocol is in the hands of the therapist.

Taken together the platforms that have been developed for the six tasks are examples of the ways in which the notions of relevance and critical subskills have been applied to define the content, context, and objectives of the part task training platform for very different complex tasks for which a VR or AR training environment has been developed. The starting point for such an effort is a training-oriented task analysis of the real-time task. In the SKILLS project this analysis capitalized on the Hierarchical Task Analysis methodology, which is routinely applied in the Human Factors domain (Duncan, 1974; Wickens et al., 2004). This method has been reoriented and reconstructed to properly cover the goals and concerns of training and skill acquisition. This reformulation was conducted by focusing on the one hand on the analysis of skill elements and training objectives, and on the other hand on the engineering implications and potential of mapping real components into virtual representations (Gavish et al., in press; Gupta and Cohen, 2002).

FACILITATION

Training systems are distinguished from unguided practice and accumulation of experience in task performance, by the inclusion of specific variables, protocols, instructions, and information that are introduced to facilitate learning and guide skill acquisition. In Chapter 2 we introduced the term *accelerators* as a global term for such facilitators. Given the wide range and flexibility in the design and introduction of such variables in multimodal VR platforms, it is instructive to review the ensemble that has been developed and experimented in the SKILLS project. Note also that although platforms were developed for very different tasks and applications, they have all been guided by a common conceptual framework and methodology. Below is a brief overview of the main accelerators studied in each platform followed by a general discussion:

Rowing: To recapitulate, the rowing platform focused on three objectives: technique optimization, improved energy management, and team coordination. For technique optimization, auditory and vibrotactile feedback were given to trainees for onset and bending of arms and back movements, in reference to expert model values. Also displayed was a visual elliptic shape of expert rowing cycle with indicators of trainee match of this shape. For energy management a human avatar on the screen signaled proportional speed change relative to the trained rower premodeled effort scale. Team coordination was trained through a requirement to adjust and synchronize movements with a virtual rower, while receiving visual feedback about the coordination discrepancies.

Juggling: Three main training accelerators were introduced to facilitate the acquisition of juggling skills in the LWJ platform. One was a speed manipulation of the global task. The manipulation changed the whole task motion and response time base from a low speed, in slow motion, to raise the speed level to normal and faster than normal levels depending on graduation criteria of competing two sequential trick cycles. Slow speed gave trainees an opportunity to observe the order and sequence of balls tossing in a trick. Note that the ability to introduce this manipulation is unique to the VR system which is relieved from gravity constraints. A second accelerator has been applied to help trainees to acquire a stable, bimanual coordinated rhythm of catching and throwing balls which typify proficient juggling. A multifrequency sensory rhythm trainer was developed combining auditory and tactile signals. The period of the sound was scaled to match the period of one cycle of one ball, while the period of the tactile pacing matched the period of one cycle of the hand. A third accelerator focused on trainees' better control of the juggling dwell ratio that is determined by the relative durations of balls in hand and in the air. Trainees were led through a sequence of juggling practice blocks in which the duration of holding a ball in hand before tossing indicated by metronome pace, or the required height of a ball throw (time in the air) as indicated by a desired height line on the LWJ screen, were systematically changed between practice blocks and displayed to trainees.

Maxillofacial: In the MFS platform, training has been focused on the acquisition of drilling competencies. The exercises required either performing a punctual drilling or drilling a line. Both are basic skills needed to do the corticotomy. Accordingly, training accelerators evolved around providing sensory information to increase sensitivity and improve discrimination between conditions. Drilling performance and learning functions were compared between different conditions of modalities' emphasis and combinations: effects of haptic-only condition; haptic, sound, and vibration combination; and visual, auditory, tactile, and force feedback joined presentation, were compared and evaluated.

Industrial maintenance and assembly: Accelerators for both the VR and the AR versions of this task are aimed at facilitating the acquisition and memorization of the multiple steps and strictly ordered procedures that have to be carried out when executing maintenance and assembly tasks on complex systems. Accelerators and training protocols were directed to assist and enhance the formation of activity sequences, on improved representations of the components and structure of the global task, as well as on facilitation of ordered retrieval in the course of execution. Accelerators have emphasized modes of task structuring and representations that can improve encoding, storage, and retrieval. Four types of accelerators were applied: observational learning—performers are presented with visual information on the steps of the tasks, to help trainees develop a mental model of the assembly/disassembly process; current action—provides direct information about the immediate action that trainees have to undertake, but only when they request it; step-level aid—provides on trainee's demand, information about a complete step and its activities; and subtask-level aid—on trainee's demand, this mode provides information about a component of the presently performed task.

Depending on the type of task and training platform (VR or AR system), information can be presented through visual pointers or graphs, text information, or the use of haptic directors.

PROGRAMMING BY DEMONSTRATION

Training operators of the programming by demonstration were directed to teach a performer to detect and avoid singularities and to better communicate compliance levels to the robot in task performance. Accelerators for avoiding singularities introduced two types of visual pointers. One has been a rotating arrow that indicates for each robot joint how close it is to the next singularity, and how to move away from it. The second provides the same information by an animated transparent robot that is showing how to move the robot elbow out of singularities. The accelerators for compliance were haptic and visual. Direct haptic feedback has been provided on parameterized levels of compliance. Operators were able to touch the robot's tool and directly feel and observe the compliance of the robot due to the forces intentionally applied by the user's hand. This multidimensional haptic aid has been expected to support the acquisition of the cognitive model of compliance through direct haptic feedback in the course of the robot movement.

UPPER-LIMB REHABILITATION

Rehabilitation is different from the other five tasks for which training platforms have been developed. While the previous five platforms were developed for healthy people with normal capabilities and advanced knowledge on the performed task, the rehabilitation platform was developed to assist therapists and equip them with new tools and power in their effort to help patients to regain motor and cognitive functionalities after impairment. Rehabilitation protocols are highly influenced by the large variability between patients in the nature and type of impairment, as well as by motivation, communication, and personality factors. Consequently, protocols have to be individually tailored and dynamically adjusted. Therefore, the primary goal of developing tasks and accelerators for rehabilitation is to take advantage of extra degrees of freedom and improved control of the robotic system in the design of tasks to be performed and the feedback given to patients. The 3D control movements afforded by the exoskeleton system and the reaching, pointing, tracking, painting, and object fitting tasks are the modular composites of constructing a therapy protocol that combines movement requirements and achievement of cognitive goals. Similarly the bimanual control system provides an inventory of tasks that call for a variety of bimanual coordinated movements and presents a variety of cognitive goals. These two toolboxes can be flexibly combined and manipulated by the therapist. An important accelerator in both systems is the ability to give a patient much more accurate and continuous feedback on their efforts and progress. This ability, which is a direct derivative of the robotic technology and improved software, empowers the therapist and the rehabilitation process over and beyond the provisions of traditional therapy.

The brief overview of the training platforms developed for the six tasks demonstrates on the one hand the wide range and richness of possible accelerators, and on the other hand the way in which task analysis, theoretical considerations, and engineering capabilities have guided the selection of a specific subset that has been developed in each platform. One important contribution of the present project has been the focus on haptics in the multimodal enaction of task-relevant perceptual motor skills. This inclusion, which capitalizes on the increasing engineering power in the design of haptic interfaces, is of special importance for tasks in which haptic-based skills are formative components of the task. It has thus been a major focus in the development and experimentation of the rowing platform for the acquisition of basic rowing skills, in which arms, legs, and upper body are coordinated in format and force operation. In the maxillofacial surgery, a key factor is haptic sensation in accurate and safe performance of drilling. The motor competencies associated with communicating compliance and avoiding singularities have been the major concern of the programming by demonstration training platform. In upper-limb rehabilitation, restoration of haptic sensation and haptic control is a prime objective of the rehabilitation process. In these four platforms, haptic competencies incorporation, proper modeling of haptic interfaces measurement, and feedback have been central in system development and testing. In the industrial maintenance and assembly system the focus has been on procedural skills, and haptic was secondary. The LWJ is an interesting case, because motor competencies empowered by haptic information are important composites of juggling with real objects. The lack of simulated haptic in this platform (resulting from engineering difficulties), led to the development and experimentation of a part task trainer in which this component is not represented. A critical test for this trainer was the transfer of training from the LWJ to juggling with real balls (see Chapter 12).

Two important categories of accelerators of training have been the provision of feedback (FB) information and knowledge of results (KR). The term *feedback* is used for information given to trainees on their own performance effort or relative to their internal reference. KR refers to information that is scaled or related to a model, optimal performance, expert behavior, or external criteria. The FB and KR aspects are separated because each may be developed and augmented separately from the other. The importance of augmented feedback is that it increases the ability of the trainee to process and retain information of his or her performance, while knowledge of results shapes the behavior of performers and drives it closer to target or model performance. The review of nine training platforms developed for the six tasks reveals a wide range of multimodal feedback indicators that were developed to guide the learning of task-related skills. It also shows in all cases how feedback and knowledge of results were integrated or related to each other.

In rowing, haptic-tactile, visual, and auditory feedback were jointly presented to inform trainees on their rowing performance. This information was also compared to expert and computational models. Effort feedback was presented by avatar speed that was scaled to both trainees' capacity and race requirements. Team coordination was informed and compared to a model displaying avatar performance. Maxillofacial demands and feedback indicators were related to models of expert surgeons' performance. Maintenance and assembly were compared to the required protocol and its accuracy. Juggling auditory and vibrotactile rhythmic pacer was associated with stable rhythm, while ball height and time in hand were related to computed stable dwell ratio. Programming by demonstration feedback and knowledge of results were driven by the model of robot motion constraints and response adjustment. Rehabilitation is driven by the functional goal of performance, and feedback is aimed to be highly sensitive and continuous to the progress and change in patient response capability. In each platform the two key questions were which aspect of performance should the feedback be developed for, and what is the standard, goal, or model to which it has to be compared.

Of special interest in the present context of multimodal training is the use of multimodal feedback indicators and in particular vibrotactile or haptic. Of the six training platforms, both rowing and maxillofacial surgery were tasks in which haptic dependent competencies were of special significance and multimodal feedback indicators combining haptic, visual, and auditory information were used. In both cases their conducted experiments showed that in order to increase the value and contribution of haptic information, task performance should be first trained under unimodal haptic feedback. Such training will increase the power of the haptic channel information as well as the benefits of multimodal over unimodal feedback of any of the three modalities by themselves (vision, audition, haptic).

Development of training environments in VR enables not only capturing and instantiating task elements and scenarios, but also augmenting reality and introducing alternative virtual situations to assist and facilitate learning. When such augmentation and virtual conditions are introduced, the key issues are the format of graduation from and the transfer of training from task performance in the virtual environment to performance in real life. Several AR elements have been introduced in the training platforms. In rowing the required rowing shape was presented by visual ellipse on the display, a human avatar was added to guide effort and team coordination. In juggling a rhythmic pacer was added. In industrial maintenance and assembly an AR presentation of the machine, the task segment, and pointers were presented on a tablet display by technician request. In programming by demonstration a virtual robot and a virtual arm were presented to indicate task steps and dangers. A key question that was coped with and answered in each of these applications is the way in which graduation and independence from the augmented information can be achieved while still benefiting from its assistance. The creation of VR that differs from natural physics is demonstrated by the speed manipulation in juggling. To enable trainees to better perceive and follow the concurrent movement of objects and their sequence, the whole task (VR world) was slowed and then gradually increased speed when trainees were able to achieve criteria performance. Here again, the key requirement is to make the best of these benefits while improving and not impairing the transfer of acquired knowledge to task performance in its natural environment.

An important category of accelerators has been the introduction and manipulation of cognitive control strategies. Cognitive control and top-down supervision are important determinants of the performance and training of every voluntary and intended task. Consequently, in many training protocols they are specifically manipulated to improve learning and influence modes of knowledge representations and its execution. In the SKILLS project the overall emphasis has been on improved enaction through the development of rich multimodal virtual training platforms (enriched bottom-up experience). The contribution and coinfluence of manipulating control strategies is of special interest. Such manipulations were included in the juggling and the industrial maintenance and assembly platforms. Within the juggling platform, trainees were given protocols in which they were asked to voluntarily change their behavior along six possible durations of holding balls in hand before tossing, or to adjust their juggling spatial trajectories such that balls are tossed to one of six indicated heights. When going through the procedures of maintenance or assembly, tasks were conceptualized at different abstraction levels and segmentation. These manipulations were added and superimposed on the normal condition of task performance in the training on the platform and compared with trainees who were not given these manipulations and practiced the tasks for equal.

TRANSFER OF TRAINING

The description and discussion of efforts to this point summarized the two major steps that have been taken in the development of training platforms for the six tasks. One was the task analysis and selection of the skills and subcomponents that will be focused upon. The other was the development and design of accelerators and training protocols. The critical test of the validity and value of these efforts is the transfer of training and the contribution of practice on the training platform to the performance of parent tasks in their real-life environment. The results of the evaluation and transfer of training studies were described in detail within the chapters dedicated to the different tasks (Chapters 11 through 16). Presented here is a global short summary of the findings and a general discussion. To recapitulate, the two aspects of transfer that should be examined are as follows: Does a VR training platform constitute a viable alternative to training time on the real task? Can the VR platform, its developed training protocols, and added accelerators enhance skill acquisition and performance competencies in aspects that are not possible or are hard to be trained in real-life conditions? The main results for the evaluation and transfer of training studies were as follows:

Rowing: Better energy management and improvement in race duration are found during transfer to a new rowing distance. Efficiency of avatar rower training for team coordination and synchronization are also found in transfer to new rowing situations.

Juggling: Light weight juggling (LWJ) platform can be an effective substitute to juggling with real balls, applying as much as equal sharing time in 10 training sessions. Training of control strategies in the LWJ improves the ability of trainees to cope with changes when juggling with real balls.

Maxillofacial surgery: Effective VR training of surgery competencies, enabling qualitative and quantitative description of major subskills and competencies constituting the MFS surgery. Systematic training and improvement are possible for the operation of surgery tools and detection of dangers.

Upper-limb rehabilitation: Overall improved progress in rehabilitation and better measurement of patient performance and progress.

Industrial maintenance and assembly: Successful replacement or preparation and assistance to real task performance with VR and AR training. Better ability with AR to avoid and correct errors.

Programming by demonstration: Better ability to avoid singularity. Improved ability to apply and adjust compliance forces.

One clear outcome of this brief overview is that there was positive transfer of training results and contributions of training in the virtual platform to the performance of the actual tasks in all six tasks. Of the two aspects of transfer, creation of effective alternative environments and benefits to skill acquisition, the ability to provide a successful alternative to training time on the real task has been supported in all cases but most directly compared in the juggling, rowing, programming by demonstration, and industrial maintenance platforms. Moreover, transfer of training comparisons on all platforms demonstrated that the especially developed accelerators and training protocols introduced in the VR platforms improved performance on aspects and competencies that are hard or cannot be included in systematic training under normal real-life conditions. The transfer of training studies has thus provided a strong support to the conceptual framework and methodologies that were experimented for the development of training platforms, as well as their accompanying engineering efforts.

CONCLUDING REMARKS

VR technologies and applications are developing rapidly and are likely to be introduced to mediate or replace direct interaction with the physical world in a growing number of daily life tasks. The richness and degrees of freedom of these technologies increase exponentially. Stimulation, monitoring, and capitalizing on these technology opportunities call for specification of domains, clarifying their goals and objectives, as well as linking the developed capabilities to existing scientific, theoretical, and experimental bases. The SKILLS project and the chapters of the book delineate the conceptual framework, methodologies, and engineering composites of developing VR training platforms for perceptual motor skills.

The driving force for this effort has been the hypothesis that it is possible to identify a significant subset of daily performed tasks in which perceptual motor competencies are a major contributor and for which efficient multimodal VR training platforms can be developed. Significant to the power and generalization prospects of the approach is the fact that the same approach concepts and methods have been successfully applied in six case studies for tasks that are very different from each other in their environment, technologies, objectives, and users. Furthermore, an important linkage to existing scientific knowledge bases had been created through the unified framework of cognitive and sensory-motor subskills. This framework served as a reference in the selection of task segments for which training is to be developed, the engineering designs and modeling efforts, as well as the development of accelerators and programming training protocols. The interdisciplinary, yet collaborative and integrative work of the 15 participating partners in this work is a good example of the important contribution and confluence of the different disciplines to the product and thinking in all stages. The work morals and general calls and challenges for the application of VR and multimodal technologies in the development of training environments have been clearly stated, specified, and experimented. They also attest to the importance of such a unified effort and collaboration in the study and application of new technologies.

REFERENCES

Duncan, K.D. (1974). Analytical techniques in training design. In E. Edwards and F.P. Lees (Eds.), *The Human Operator in Process Control* (pp. 283–319). London: Taylor and Francis.

Gavish, N., Krupenia, S., and Gopher, D. (In press). Task analysis for developing maintenance and assembly virtual reality training simulators. *Ergonomics in Design.*

Gopher, D., Krupenia, S., and Gavish, N. (2010). Skill training in multimodal virtual environment. In D. Kaber and A. Boy (Eds.), *Advances in Cognitive Ergonomics* (pp. 884–893). London: Taylor and Francis.

Gupta, P., and Cohen, N.J. (2002). Theoretical and computational analysis of skill learning, repetition priming, and procedural memory. *Psychological Review,* 109, 401–448.

Sanchez-Vives, M.V., and Slater, M. (2005). From presence to consciousness through virtual reality. *Nature Reviews Neuroscience,* 6, 332–339.

Slater, M., Lotto, B., Arnold, M.M., and Sanchez-Vives, M.V. (2009). How we experience immersive virtual environments: The concept of presence and its measurement. *Anuario de Psicología* 40, 2, 193–210.

Wickens, C.D., Lee, J.D., Liu, Y., and Gordon-Becker, S.E. (2004). *An Introduction to Human Factors Engineering* (2nd edition). Upper Saddle River, NJ: Pearson.

18 Future Perspectives and Technological Challenges in Multimodal Human-Machine Interaction

Massimo Bergamasco

CONTENTS

INTRODUCTION

The new approach followed in SKILLS for the training of a human operator in different application domains, as described in the previous chapters, has determined also an innovative approach to the design of the Virtual Environments systems utilized in the framework of the different demonstrators.

The main innovations in terms of functionalities introduced by SKILLS in a virtual environment (VE) system can be shortly summarized as follows:

1. The objective of generating new complete "experiences" for the human operator instead of replicating the same operations performed in a real environment; such interaction experiences can be focalized to specific subskills (both cognitive and sensorimotor) or components of the skill to be acquired.
2. The introduction of "accelerators" in the control loop of the interaction process. Such accelerators refer to the implementation of a metaphor of the task

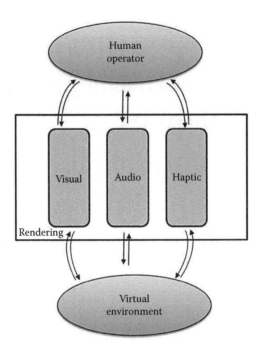

FIGURE 18.1   A block diagram of the interaction process between the human operator and the virtual environment.

   (i.e., a reduction in complexity or focus on a single or a few aspects of the task to be trained).
3. The introduction of a digital representation component aimed at detecting and modeling (and even replicating to an external robotic or virtual agent) the characteristics of a specific act of motion performed by the human operator.

The interaction process between the human operator and the VE can be represented as in Figure 18.1, where the human operator experience is completely mediated by the multimodal interface. Each sensory pathway (visual, auditory, and haptics) presents a bidirectional flow of information (efferent and afferent data) between the human operator and the VE. Each sensory modality is handled by the VE controller that generates appropriate stimuli at the natural frequency rate for the specific modality (e.g., 1 KHz for the haptic pathway, 60 Hz for the visual pathway).

   The natural extension of the interaction process with VE integrates real-world elements into the perceptual experience of the human operator. According to the concept of mixed-reality continuum (Milgram and Kishino, 1994), augmented reality (AR) is considered as the condition in which some VE elements are integrated in a real environment representation. Despite the technological thresholds introduced by AR, a smoother and natural interaction process can be achieved in this condition.

   Figure 18.2 presents a possible arrangement of the block diagrams for an AR system. A first feature of such a solution is that the capturing process is not limited

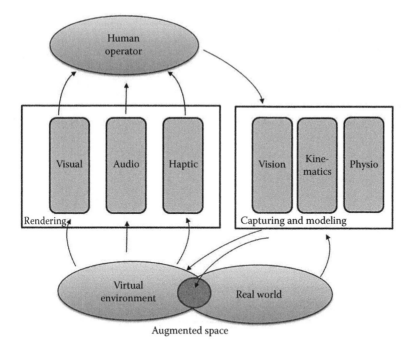

FIGURE 18.2 A block diagram of the interaction process between the human operator and the augmented space.

to the efferent flow of the human operator, but it takes into consideration also the objects or entities belonging to the real environment. Correspondingly, the entities of the real world are modeled in the VE in order to generate a coherent interaction with the virtual entities (e.g., for casting shadows and occlusions or for the physical interaction with virtual objects).

A relevant example of AR integration in SKILLS is presented in Webel et al. (2011) and Ruffaldi et al. (2011).

VIRTUAL ENVIRONMENTS IN SKILLS

A common, and more specific representation of the interaction with a VE/AR system is shown in Figure 18.3, where the two efferent (Capture) and afferent (Renderers) blocks of the interface system are separated.

The interaction process implemented in different demonstrators of the SKILLS project has been modified by adding four elements:

1. A capturing phase. The modeling of the user's performance is based on the capturing phase executed in the real world or on the VE platform by taking into consideration both expert and novel users.
2. A digital model of performance, instantiated from the captured data and able to provide performance indicators in real time.

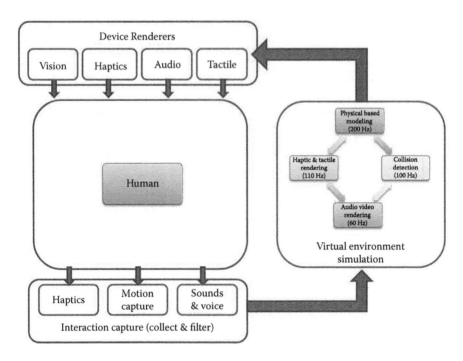

FIGURE 18.3   Detailed representation of the control loop for a common VE system.

3. A set of training accelerators that can be considered specific software components that provide the human operator with appropriate feedback for the acquisition of a specific component of skill.
4. A training manager which is a software component able to steer the training protocol by exploiting the performance indicators generated by the digital model of performance, and by activating the accelerators.

The four innovative elements introduced in a VE system are presented in Figure 18.4 in the specific case of the rowing demonstrator (see Chapter 11).

By taking into account the four components introduced into a VE system according to the SKILLS approach, the control loop shown in Figure 18.3 can be extended as reported in Figure 18.5.

One of the most important components for the future developments of VE systems for training is represented by the digital representation (DR) of skill. Figure 18.6 depicts the arrangement of the conceptual blocks involved in the creation of DR. In the task space several operations are performed on the captured data in order to extract a model of expert performance (segmentation, labeling, classification, and regression). In addition, the understanding of a subset of internal states of the subject (physiological state, intentional and motivational aspects) represents an important element of the DR modeling.

The role of the DR model can be extended to the control of virtual or robotics agents in order to achieve the embodiment of an expert performance into a synthetic

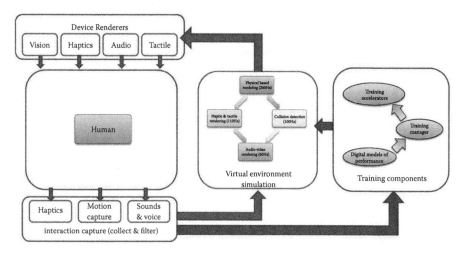

FIGURE 18.4 The innovative components of a VE system in SKILLS as applied in the rowing demonstrator.

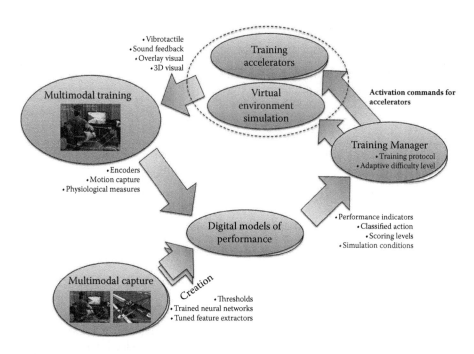

FIGURE 18.5 Detailed representation for a SKILLS VE system.

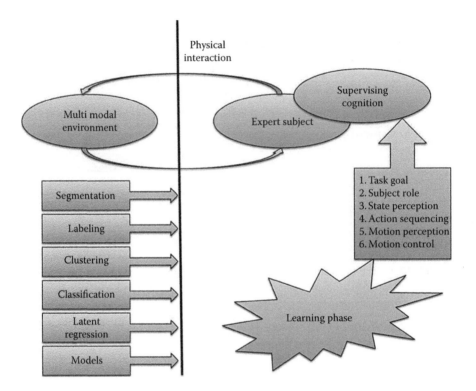

FIGURE 18.6 Conceptual blocks of the learning phase for creation of the digital representation.

entity (Figure 18.7). Such an approach could be effective in the future to extend the training of rowing teams as discussed in Chapter 11.

FUTURE PERSPECTIVES

Given the final stage of research development of the SKILLS project, three main research areas can be identified where the corresponding technological challenges will assume paramount importance in the future.

PERSPECTIVES IN DATA ACQUISITION AND PROCESSING

The possibility of acquiring more and better organized data, and at the same time, the application of more sophisticated analysis techniques are two challenges for the acquisition and processing of human skills. There are several technological advancements that can overcome the criticalities posed by these challenges. From one side there is an increased availability of wearable sensors whose physical size is being reduced together with improvements in power consumption and integration. The miniaturization push from the domain of smartphones has indeed made available smaller sensors for other application domains with an associated reduction of costs. At the same time there is an increase in computing power both at the local portable

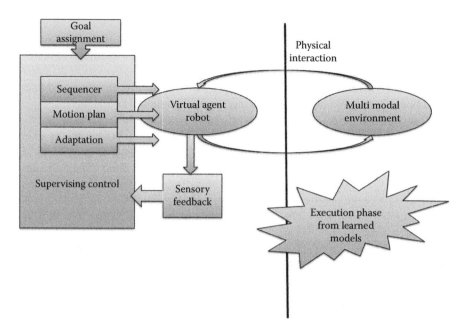

FIGURE 18.7 Conceptual block of the application of the digital representation (DR) model to a virtual agent. Specific importance is given to the supervising component that plans and executes the DR model.

level and at the networked level by means of the diffusion of cloud computing platforms. Finally, the improvements in wireless network connectivity are making devices and sensors more portable and flexible in their usage. Given these technological advancements, it is possible to identify several directions for the evolution of the data management associated with skill data processing.

The first direction is the possibility of performing advanced capturing and modeling directly on wearable systems allowing results and feedback of user performance to be synthesized in place. In case the local computing power is not sufficient, an already established solution is the possibility of employing large resources from cloud computing for synthesizing results or performing recognition based on large corpora, provided that the input information is compatible with the bandwidth of the network. Recent examples are from the domain of speech recognition and landmark visual recognition: the wearable sensor captures the information and encodes it in a compact representation with the distinctive features of the original signal. Then the expensive recognition task is performed remotely on the cloud. With the increase in local computing power and network bandwidth, this scheme will be generally useful when the recognition requires the use of large datasets or whenever for security reasons it is not advisable to store the dataset on the local device. The increase in computing resources is supporting the adoption of more data-intensive algorithms.

There is a large interest in the creation of large datasets that can be used for the extraction of common models and later use for user training and evaluation. These models rely on the evolution in machine learning techniques supported by the large computing resources on the cloud. These corpora of user performance are currently used for training motion and action recognition systems (Shotton et al., 2011) or for building motion libraries for supporting animation. There are few examples of multimodal captures of user performance, but they do not span multiple users or different expertise levels (Tenorth et al., 2009). Data intensive means also that acquired data should be ready for processing with minimal manual labeling or curation.

The second direction is the possibility of obtaining flexibility in the sensing system by means of easy wireless integration of personal sensors. These sensors span from motion sensors for the different body parts to ones supporting physiological measures. The reduction in size, connectivity, and distributed computing will support the extraction of this information and its integration. In particular there is a trend toward zero-power or passive sensors (Ionescu and Hierold, 2011), while another is the miniaturization for the embedding of sensing elements in cloths or in flexible structures in general.

The natural extension of personal sensors is the extension to teams that are monitored and trained together. In this way it would be possible to have an overall view of team performance and an integrated feedback. The challenges for these results are both at the level of the sensor network in flexible environments and the architecture for the associated data (Hoyt, 2008).

Complementary to personal sensors is the integration of environment sensors that communicate with the user in a flexible way. This means that an environment could provide a motion capture service while the user is passing nearby a room, and this could provide additional information to embedded skill feedback systems. The creation of environments that provide sensors as services to various users depends mostly on standardization and on the definition of proper software interfaces.

In terms of theory of machine learning and its application, there are two major challenges. The first is related to the selection of the best and most effective methods given the data, due to the fact that there are many available techniques for performing the same operation. The second is instead related to extracting a structure from a complex and large dataset, an aspect that has been tackled by means of hashing and sparse representations from one side, and by means of deep networks from another.

Future Directions in Digital Representation

The achievement reached in the SKILLS project in terms of digital representation, not only helped to define new tools and algorithms for the management and modeling of sensory-motor data in multimodal training, but also assisted in establishing a common ground between motion modeling and synthesizing, interaction representation, action benchmarking, and training acceleration.

Even if within the SKILLS framework those technologies have been applied to a reduced set of demonstrators, the achieved digital model showed an intrinsic ability to mediate between the control actions and the environmental changes that are perceived through sensors. This feature shows high potentialities in different fields which are well beyond the scope of the SKILLS project, such as domestic robotics, dexterous and skilled manipulation, complex skills learning and composition, and expert benchmarking.

The incremental learning and the sensory-motor relationships caught in the skills representation offer to future engineers an opportunity to overcome the unstructured environment barrier that prevents most robots from having the capability of operating in social populated (and even working) environments.

AUGMENTED REALITY AND SKILL TRANSFER

Throughout this book we have seen how the use of VE technology and methods has proven to be of critical importance on the task of acquiring, studying, and transferring skills, but the future ahead is even more promising as a new wave of technologies and research results are achieved in the field of AR. New display technologies such as the holographic optical waveguide (Cameron, 2009) are about to debut on low-cost commercial products, finally allowing for light, efficient, and wide Field of View (FOV) AR glasses, a critical element to bringing information visualization directly on the field, overlaying computer-generated images to the very same subject under study, or to the surroundings where our training is taking place. This is a key element to provide visual information with a clear and understandable relationship with the real-world objects it is related to, effectively addressing issues due to communication (language barriers, misunderstandings, etc.), attention, comprehension, and memory.

Tracking technologies and methods keep their ineluctable progress toward the grail of AR: generic, unconstrained, stable, drift-free localization and tracking. The new generation of inertia sensors and dead reckoning methods are promising for drift-free outdoor and indoor localization (Foxlin, 2005), while computer vision Simultaneous Localization And Mapping (SLAM) is reaching that level of sophistication (Castle et al., 2011) where large un-instrumented areas can be covered with a sufficient level of precision to guarantee the colocation between the real-world and the augmented information.

Finally, wearable computing, another essential element for the successful deployment of AR, is taking advantage of the innovations of low-power computing, a sector that is attracting huge investments due to the emerging market of smart mobile phones and tablet computers. Month after month we are witnessing a new generation of low-power architectures (the current forerunner appear to be ARM architectures), fully capable of handling 3D graphics with the performance equivalent to office workstations of just 5 years ago (NVIDIA Corporation white paper, 2011). These achievements will not only lead to offer more appealing, realistic, structured, and immersive content, but will also result in a wider and more pervasive use of these technologies, as confirmed by the current trends.

All these converging elements truly represent high-impact enabling technologies and are likely to produce a surge of the use of AR in several training and learning scenarios. When the technological constraints will be less crucial, the attention will probably shift to content providing; the challenge will then become to find methodologies and cost-effective tools in order to produce, reference, access, and share even larger amounts of digital information, opportunely assessed and validated, in a reasonable time.

CONCLUSIONS

Multimodal interfaces represent an essential component for the future development of the interaction process with VE and AR. The new trends in information and communication technology already demonstrate that computing power and graphical capabilities can evolve toward solutions possessing a high degree of pervasiveness. SKILLS introduced a new approach in the design of future training applications, with innovative theoretical achievements in the field of digital representation of skill. Technologies of human-machine interfaces together with new data mining techniques will bring the concept of future training toward more personalized and ubiquitous solutions.

ACKNOWLEDGMENTS

The author wishes to thank Emanuele Ruffaldi, Carlo Alberto Avizzano, Franco Tecchia, Marcello Carrozzino, and Federico Vanni for the fruitful discussions on the future vision of VE in training and their help in the definition of new concepts reported in this chapter.

REFERENCES

Cameron, A.A. (2009). The application of holographic optical waveguide technology to the Q-Sight family of helmet-mounted displays, *Proc. of the SPIE* 7326, 7326H.

Castle, R.O., Klein, G., and Murray, D.W. (2011). Wide-area augmented reality using camera tracking and mapping in multiple regions, *Computer Vision and Image Understanding*, 115, 6, 854–867.

Foxlin, E. (2005). Pedestrian tracking with shoe-mounted inertial sensors, *IEEE Comput. Graph. Appl.*, 25(6): 38–46.

Hoyt, R.W. (2008). SPARNET–Spartan Data Network for Real-Time Physiological Status Monitoring, Defense Technical Information Center (DTIC) Document, Fort Belvoir, VA.

Ionescu, A.M., and Hierold, C. (2011). Guardian angels for a smarter life: Enabling a zero-power technological platform for autonomous smart systems, *Procedia Computer Science*, 7, 43–46.

Milgram, P., and Kishino, F. (1994). A taxonomy of mixed reality visual displays, *IEICE Transactions on Information and Systems*, 77(12): 1321–1329.

NVIDIA Corporation, White Paper: Variable SMP—A Multi-Core CPU Architecture for Low Power and High Performance, retrieved at http://www.nvidia.com/content/PDF/tegra_white_papers/tegra-whitepaper-0911b.pdf (accessed January 13, 2012).

Ruffaldi, E., Tripicchio, P., Avizzano, C.A., and Bergamasco, M. (2011). Haptic rendering of juggling with encountered haptic interfaces. PRESENCE 20-5, Cambridge, MA: MIT Press.

Shotton, J., Fitzgibbon, A., Cook, M., Sharp, T., Finocchio, M., Moore, R., Kipman, A., and Blake, A. (2011). Real-time human pose recognition in parts from single depth images, *CVPR*, 2(3). Kyoto, Japan: IEEE.

Tenorth, M., Bandouch, J., and Beetz, M. (2009). The TUM kitchen data set of every-day manipulation activities for motion tracking and action recognition, *Computer Vision Workshops (ICCV Workshops), 2009 IEEE 12th International Conference on*, pp. 1089–1096.

Webel, S., Bockholt, U., Engelke, T., Gavish, N., and Tecchia, F. (2011). Design recommen-dations for augmented reality based training of maintenance skills. In *Recent Trends of Mobile Collaborative Augmented Reality Systems*, ed. Leila Alem et al., 69–82. New York: Springer.

Index

Printed and bound by CPI Group (UK) Ltd, Croydon, CR0 4YY

18/10/2024

01776262-0010